AF168360

Future of Business and Finance

The Future of Business and Finance book series features professional works aimed at defining, describing and charting the future trends in these fields. The focus is mainly on strategic directions, technological advances, challenges and solutions which may affect the way we do business tomorrow, including the future of sustainability and governance practices. Mainly written by practitioners, consultants and academic thinkers, the books are intended to spark and inform further discussions and developments.

Mayank Kejriwal

Artificial Intelligence for Industries of the Future

Beyond Facebook, Amazon, Microsoft and Google

 Springer

Mayank Kejriwal
Information Sciences Institute, Ste 1001
University of Southern California
Marina Del Rey, CA, USA

ISSN 2662-2467 ISSN 2662-2475 (electronic)
Future of Business and Finance
ISBN 978-3-031-19038-4 ISBN 978-3-031-19039-1 (eBook)
https://doi.org/10.1007/978-3-031-19039-1

© The Editor(s) (if applicable) and The Author(s), under exclusive license to Springer Nature Switzerland AG 2023
This work is subject to copyright. All rights are solely and exclusively licensed by the Publisher, whether the whole or part of the material is concerned, specifically the rights of translation, reprinting, reuse of illustrations, recitation, broadcasting, reproduction on microfilms or in any other physical way, and transmission or information storage and retrieval, electronic adaptation, computer software, or by similar or dissimilar methodology now known or hereafter developed.
The use of general descriptive names, registered names, trademarks, service marks, etc. in this publication does not imply, even in the absence of a specific statement, that such names are exempt from the relevant protective laws and regulations and therefore free for general use.
The publisher, the authors, and the editors are safe to assume that the advice and information in this book are believed to be true and accurate at the date of publication. Neither the publisher nor the authors or the editors give a warranty, expressed or implied, with respect to the material contained herein or for any errors or omissions that may have been made. The publisher remains neutral with regard to jurisdictional claims in published maps and institutional affiliations.

This Springer imprint is published by the registered company Springer Nature Switzerland AG
The registered company address is: Gewerbestrasse 11, 6330 Cham, Switzerland

To the future generation of my family: my niece, Mishka, and my nephew, Atharv.

Preface

Everyone has a view on artificial intelligence (AI) these days: scientists, policy makers, lawyers, engineers, journalists, consultants, and even artists. For a relatively young researcher in the field, this may seem like a welcome sign: after all, isn't public and government interest a sure-shot sign, as any, of increased funding and opportunities?

While this excitement can be heady, it has the unfortunate consequence of subjecting the field and its advancement to a degree of speculation that is not always supported by the evidence. I do not dispute that speculation, and its more creative cousin, imagination, can have intrinsic value; they prepare us for what might lie ahead, and they can be sources of inspiration for new ideas and modes of thought. At the same time, for both industry and government to make sound decisions and rationally allocate resources, it is important to be able to separate fact-based trends from pure speculation, and to also think about what the most reasonable outcomes might be, even if only the present state of the technology was taken into account.

The goal of this book is to set out the role of AI in "industries of the future." It is not to speculate on "what" those industries might be, since history has taught us to be humble about our abilities to predict inventions in the long-term future. Who could reasonably have predicted, even in 2000, that a company like Apple would become a trillion-dollar (and counting) future smartphone maker or that commercial space tourism, pioneered by a start up, would soon be on the horizon? Not to mention the normalization of remote work, uniquely twenty-first century problems such as social media addiction, a decentralized cryptocurrency (Bitcoin) that takes the energy of a small country to "mine," and an AI-composed artwork that has spurred international debate on whether an AI can be granted copyright on its work (Chap. 6)? In hindsight, it seems that we may have better luck (which is to say, very little) predicting business cycles than in predicting the path, let alone the impact, of advanced technology.

I would hope that my aim is less ambitious, but no less useful. Put simply, I am not going to spend much time speculating about *what* the inventions or industries of the future will be. Nor will I be discussing the projected impacts of technology that *currently* does not exist (and may not exist for another twenty years, or, for all we know, could be mainstream only a few years from now) might have. Instead, this book is predicated on the assumption that AI, as it exists today, already allows us to say something important about potential industries of the future. In other words, we do not have to (necessarily) know what future AI will look like, any more than we need to try and define markets that do not exist yet, to still be able to *meaningfully* speak about what industry today should be thinking about as they set their sights toward an uncertain future.

What might some such features be? First, press and media hype aside, the sobering reality is that many companies are still a long way off from implementing AI today. In part, one issue is that measuring the return on investment (ROI) of AI is challenging, if not methodologically unprecedented. In Chap. 2, I go deeper into this issue after first introducing AI in Chap. 1, clarifying some of the terminological choices that I made when planning and writing this book. Fortunately, industry watchers and consultants have already thought about the thorny issue of AI implementation and ROI in some detail, guided undoubtedly by the operational, strategic, and pragmatic concerns and challenges that they have personally encountered in their conversations with clients and company leaders. My efforts in that chapter are to synthesize their lessons in a way that can be availed of (at least as a starting point) by any company looking to implement AI more vigorously throughout the fabric of their organization.

Even if we can't predict with great specificity what industries of the future might be, we can reasonably infer where they may emerge from. I argue that there are three important (and rather unsurprising) such sources: Big Tech, start-ups, and major corporations that are not Big Tech, but that are looking heavily to emerging technologies to improve their productivity and future growth prospects. In Chap. 3, I consider all three groups, and show that, in looking at the current advances being made in each, it is possible to say something (albeit with caution) about new AI-driven products and services that are already on the horizon, and could be impressive money-makers for the companies furthest along in developing and commercializing them. One such field of AI that I focus on as a mini-case study is Natural Language Processing (NLP) where truly remarkable progress has been achieved in both industry and academia over the last half decade. While much of the focus is on the United States, I also briefly discuss the important role of the Chinese Big Tech companies, many of which are also at the bleeding-edge of research in high-impact areas like Web search, NLP, and computer vision.

Irrespective of who the winners and losers will be in this race, whether nationally or internationally, there is no question anymore that both the government and human resources divisions within corporations will have to be involved. Chapter 4 considers how the workforce will be affected by "augmented" AI, which (in contrast with "automated" AI) assumes that AI is an advanced tool or implement that must work *alongside* humans to solve complex problems. Most likely, this is the variety of AI that will get implemented in practice in large organizations. This is not to say that every job or occupation will be equally affected. Similar to effects of technologies in the past (including the personal computer and word processing software), some jobs will get automated away, and additionally, some sectors will be more heavily affected than others. Similar to Chap. 3, I aim to use actual examples and consulting studies to show what these effects are likely to be. I present a small case study of a medical sub-field (radiology) that we may not think is at risk of upheaval by AI; in fact, I draw on scholarly writing by actual radiologists to show that the changes are already happening, and radiologists need to adapt to the reality that AI is already quite good at (if not better than) some of the core tasks that radiologists are called upon to do today.

Chapter 5 dives deeper into AI ethics and policy. This is a rapidly evolving area, with each year bringing new developments to the fore. However, some recently passed laws and trends that are relevant to both AI and privacy (such as the European General Data Protection Regulation or the GDPR) are likely to stay, and we are already starting to see some of their effects. No company with a sufficiently large market size, or technological footprint, can afford to ignore these regulations and trends.

Chapter 6, wherein I conclude the book, is more future-facing than the others. I provide some guidance on where the future is headed. Once again, I try to avoid speculation by tying each issue that I discuss into ongoing events that provide an evidence-based foundation for why I included the issue (in that chapter) in the first place. One example (also mentioned earlier) is AI copyright, which, as futuristic as it might sound, is already being debated in patent and copyright agencies in several countries as part of a global project currently being led by a law graduate.

At this juncture, I would like to clarify my use of single quotes and double quotes. The former is used when I am drawing attention to a term, the usage of which may be less obvious or more questionable than it first seems, while the latter is always used when quoting directly from a source. Occasionally, I also used italicization and bold text to draw attention to a phrase, term, or its contextual usage.

To conclude, we are at an exciting juncture in our technological journey as a species. We may not know what is to come, or what specific form it will take, but current AI technology has advanced far enough as to enable us to make some reasonable claims about industries of the future. Although short, this book draws on a mix of scholarly sources, media reporting, and consulting pieces and commentary, to synthesize those claims in a single work. My hope is that it will prove to be useful to industry leaders and other interested stakeholders, both as an accessible review of contemporary perspectives on AI's forward-looking role in industry as well as a clarifying guide on the major issues that companies are likely to face as they commence on this exciting path.

Marina Del Rey, CA, USA   Mayank Kejriwal
July 2022

Acknowledgments

I have no hesitation in acknowledging that this book is built largely on the knowledge of others, but the most important contributor may be self-reporting by industry itself. In particular, a number of consulting firms have been investigating successful integration (or the possibility thereof) of AI in companies, including best practices and return on investment (ROI), and that knowledge has been disseminated regularly through whitepapers and reports. I have drawn upon that reporting extensively, with credit and citations accorded where due, for certain segments of this book, as their conclusions have been guided by actual (rather than academic or even hypothetical) problems encountered when working with clients. Certainly, without that dissemination, this book would not be possible. Credit also goes to the greater degree of openness (at least concerning AI research) with which many companies, both within Big Tech but also startups, are operating today. Open-source projects are now common in many industrial sectors, as is academic publishing and presentation of the best work of the researchers working in those companies. This openness allows the book to be non-speculative, and to focus on reality than on counterfactuals or hypotheticals.

My acknowledgments also extend to the growing body of commentators and journalists in the media who have increasingly started to report on AI-related issues, including political and ethical issues. Their coverage has informed many aspects of my own, and the book attempts to credit them by citing news reports where relevant. Some may argue that this makes the book less "scholarly" than a book that is more traditionally academic; however, I believe that, for a rapidly evolving topic that is focused on industry as a first-class citizen, reporting from reputed news sources as well as discussions in industry whitepapers do constitute a primary source of material.

Last, but not least, I credit policy research done (and disseminated) by governments and policy makers that have informed certain chapters in the book. Such research is usually conducted by advisory bodies and special councils comprising experts who often (voluntarily, and for free) lend their knowledge to advising governments and lawmakers on best practices for regulating and adopting AI. Their thought leadership informs this book, which is, ultimately, a mere synthesis and synopsis of their more extensive and original explorations.

Contents

Acronyms

AGI	Artificial General Intelligence
AI	Artificial Intelligence
AI2	Allen Institute for Artificial Intelligence
AICT	Artificial Intelligence Capabilities and Transparency Act
AIM	Artificial Intelligence for the Military Act
API	Application Programming Interface
AWS	Amazon Web Services
BATX	Baidu, Alibaba, Tencent, and Xiaomi
BERT	Bidirectional Encoder Representations from Transformers
CASMI	Center for Advancing Safety of Machine Intelligence
CASP	Critical Assessment of Techniques for Protein Structure Prediction
CoE	Council of Europe
CSR	Common Sense Reasoning
DABUS	Device for the Autonomous Bootstrapping of Unified Sentience
DARPA	Defense Advanced Research Projects Agency
DL	Deep Learning
DoD	Department of Defense
DQD	Data Quality Dimension
EBITDA	Earnings Before Interest, Tax, Depreciation and Amortization
EHR	Electronic Health Record
EU	European Union
EY	Ernst & Young
FANG	Facebook, Amazon, Netflix, and Google
FAANG	Facebook, Amazon, Apple, Netflix, and Google
FDA	Food and Drug Administration
FTC	Federal Trade Commission
GAN	Generative Adversarial Network
GDP	Gross Domestic Product
GDPR	General Data Protection Regulation
GovAI	Centre for the Governance of AI
GPAI	Global Partnership on Artificial Intelligence
GPT	Generative Pre-Trained Transformer
GPU	Graphical Processing Units
HAI	Human-Centered AI

HEART	Hydrogen Electric and Automated Regional Transportation
HR	Human Resources
IDE	Integrated Development Environment
IE	Information Extraction
i.i.d	independent and identically distributed
IoT	Internet of Things
IP	Intellectual Property
IT	Information Technology
JEDI	Joint Enterprise Defense Infrastructure
KB	Knowledge Base
KG	Knowledge Graph
KPMG	Klynveld Peat Marwick Goerdele
LM	Language Model
LRM	Language Representation Model
LORELEI	Low Resource Languages for Emergent Incidents
LwLL	Learning with Less Labeling
M&A	Mergers & Acquisitions
MAMAA	Microsoft, Apple, Meta, Alphabet, and Amazon
MCS	Machine Common Sense
ML	Machine Learning
MOOC	Massive Open Online Course
MRI	Magnetic Resonance Imaging
NDAA	National Defense Authorization Act
NIST	National Institute of Standards and Technology
NLP	Natural Language Processing
NLU	Natural Language Understanding
NPV	Net Present Value
NPR	National Public Radio
NSCAI	National Security Commission on Artificial Intelligence
NSF	National Science Foundation
NYT	The New York Times
NYU	New York University
OECD	Organisation for Economic Co-operation and Development
PaLM	Pathways Language Model
PDSS	Personalized Decision Support Systems
PwC	PricewaterhouseCoopers
QIS	Quantum Information Science
R&D	Research & Development
RDBMS	Relational Database Management System
RL	Reinforcement Learning
ROI	Return on Investment
RPA	Robotic Process Automation
SAIL-ON	Science of Artificial Intelligence and Learning for Open-world Novelty
SBIR	Small Business Innovation Research

SEO	Search Engine Optimization
SME	Small and Medium-Sized Enterprise
STEM	Science, Technology, Engineering, and Mathematics
STTR	Small Business Technology Transfer
SVM	Support Vector Machine
TFQ	TensorFlow Quantum
TPU	Tensor Processing Units
UK	United Kingdom
UN	United Nations
UNESCO	United Nations Educational, Scientific and Cultural Organization
UNICRI	United Nations Interregional Crime and Justice Research Institute
US	United States
USA	United States of America
USPTO	United States Patent and Trademark Office
VLSI	Very-Large Scale Integration
WIPO	World Intellectual Property Organization
WSJ	The Wall Street Journal

Artificial Intelligence: An Introduction

1

1.1 Introduction

Artificial Intelligence (AI) has always been a popular subject in pop culture, science fiction, and discussions of the future, but press and commentary over the last decade (in particular) has made the future seem a lot closer. On typing "artificial intelligence" into Google News (as the author did, on June 20, 2022), articles are returned from the likes of VentureBeat, Forbes, Scientific American, ABC News, and even the US Patent and Trademark Office (USPTO). Here are some headline examples that can be interpreted as being "bullish" on AI being rolled into our lives, some much wilder than others:[1]

- CNN: *Is Artificial Intelligence getting too human for comfort?*
- InformationAge: *5G and AI use cases—how 5G lifts artificial intelligence?*
- euronews.next: *How Artificial Intelligence is revolutionizing healthcare*
- United States Patent and Trademark Office: *Artificial Intelligence for all*
- Times of India: *"Artificial Intelligence is rewriting the game"*
- ZDNet: *Artificial intelligence projects grew tenfold over past year, survey says*
- Giant Freakin Robot: *Google is close to achieving true Artificial Intelligence?*
- Yahoo Finance: *Global Medical Imaging Equipment Market Report 2022: A $48.58 Billion Market in 2026—Integration of Artificial Intelligence (AI) with Medical Imaging Equipment Gaining Traction*
- Power Technology: *Artificial Intelligence companies leading the way in the power industry*

---

[1] As is all too common in online news today, the title can sometimes be more dramatic than the actual content of many such articles, which tend to be more moderate in their conclusions.

© The Author(s), under exclusive license to Springer Nature Switzerland AG 2023
M. Kejriwal, *Artificial Intelligence for Industries of the Future*, Future of Business and Finance, https://doi.org/10.1007/978-3-031-19039-1_1

At the same time, the press is doing a better job than in a previous era of recognizing that AI may be going through another hype cycle. Here are some headline examples from the same Google News search that express some healthy skepticism:

- Scientific American: *Artificial General Intelligence is not as imminent as you might think*
- VentureBeat: *"Sentient" artificial intelligence: Have we reached peak AI hype?*
- Times of Malta: *Myth debunked: Artificial Intelligence will replace data scientists*
- Mother Jones: *The long, hype-strewn road to general Artificial Intelligence*

While some of this commentary is (unfortunately) designed to be hyped and lull an unsuspecting reader into the article as click-bait, much of the speculation on where AI is headed is not entirely without cause, and some of it may be surprisingly accurate. This is especially the case when projections are made for specific industries based on real discoveries that are being rolled out into products and services. Radiology, predictive maintenance and the power industry are all good examples, and we provide a case study on how AI might change radiology's value proposition in a subsequent chapter. More broadly, in a post-COVID world, organizations have become all too aware of how quickly their workers may be able to transition to new technologies and ways of working. If anything, we may be underestimating how quickly some sectors of the economy could change. Change here should be thought of as being more comprehensive and nuanced than the usual theatrical trope of robots (or less excitingly, computer software) slowly replacing warm human bodies. Even without such radical automation, technological advances such as the ability to hold a fairly sophisticated computing and communication device (our smartphones) allow us to be ubiquitously online, handle meetings and communications from anywhere in the world, and with the added modern benefit of cloud computing, run experiments, and delegate major data crunching tasks "to the cloud" through a laptop with an Internet connection.

There is very real evidence that, partly because of technology and (post-COVID) acceptance of remote work, implementing a 4-day workweek without loss of productivity may be a pill that organizations are willing to swallow. In the United Kingdom, for example, a massive trial has just begin where thousands of British workers employed at voluntarily participating employers have adopted the essence of a 4-day workweek for the last six months of 2022 [28]. One of the aims of the trial is to gain a more scientific and data-driven (as opposed to speculative) understanding of whether such a model might serve as the basis for the future of work. An informal consequence may be that moves such as these may make the notion of a 4-day workweek more mainstream and eventually lead to cultural and social acceptance.

Without AI and technology, it is doubtful that the idea that a 4-day workweek could be just as (and counter-intuitively, may even turn out to be *more*) productive as a 5-day workweek, would have been seriously entertained. Note that the fact that the trial is taking place now is a testament to how far technology has already come.

In other words, we may not have to wait for an AI revolution of sorts before it starts to make its impacts felt in the workplace in the here and now. In Chap. 2, we provide guidance on how companies could go about thinking about the return on investment (ROI) of AI projects and why the devil is so often in the details. In Chap. 4, we also discuss how AI could augment the workforce and bring about changes in the way that we view work.

However, before we can start commenting on AI and its effects on industries of the future, it is important to clarify just what AI is and also what we mean by *industries of the future*. These are the goals of the present chapter. We begin by introducing the basics of AI for contextualizing the remainder of the book. Our aim here is not to provide a thorough technical description of AI, since many established textbooks on AI already undertake such a task, but to provide enough of a foundation for the reader not well versed in AI. We pay special attention to the relationship between AI, machine learning, and deep learning, which tend to be conflated in many discussions.

Concerning industries of the future, we first survey the definitional landscape of *Industry 4.0* (a better known term), since different people can often mean different things when referring to such terms ,, not unlike AI. However, we note at the outset that, although industries of the future, as we conceive it, cannot be discussed without reference to Industry 4.0, the two are not equivalent. Beginning with a sufficiently comprehensive description of Industry 4.0, we comment on why we believe this to be the case. Finally, we end the chapter with a discussion on where or how industries of the future could emerge, and why AI is critical to the entire movement. The rest of the book dives deeper into each of these themes, which this opening chapter only seeks to introduce.

1.2   Artificial Intelligence (AI)

To understand AI and what its boundaries are, it is useful to take a brief trip back in time. The idea of automated machines being able to engage in challenging human endeavors, such as playing chess, has always been a fascination, as evidenced by the great hype accorded to the "Mechanical Turk" machine constructed in the late eighteenth century (and later revealed to be an elaborate hoax, with a human chess-master operating the machine from the inside) [3]. However, a serious discussion on computing and intelligent machines began in earnest with the pioneering writings and work of Alan Turing (since made famous in pop culture through the modern movie "The Imitation Game") and became academically mainstream in the mid-late 1950s [19]. Indeed, by that time, a community of AI researchers and logicians had emerged, and arguably, the first prototypes of AI software started crystallizing. The influence of some of these continues to be felt till this day, e.g., the ELIZA chatbot in the mid-1960s [33]. Automated reasoning was already witnessing rapid advances for a scientific area in that period (including a very early version of modern neural networks, called the perceptron [24], in the late 1950s), although it has not since been a smooth and continuous journey since then. There have been a number of "AI winters" along the way [15], where a new generation of AI concepts and findings were initially thought to be the key to intelligence, only for their limitations to become apparent when they were put through human-like challenges.

Today, the situation is very different; yet, there is some fear that another AI winter may well be on the horizon (or at minimum, that we may be exiting an AI "summer") [14]. In this book, we are agnostic to such perspectives. Instead, we consider AI as a long-term trend that is changing industry. In some sectors, this adoption is going to be slower than others and not always because of the companies themselves. In domains where human life is on the line, for instance, one would always expect a higher barrier to entry and more stringent performance requirements. Systems for achieving such performance targets may take longer to develop than others. Our thesis is that, regardless of the messiness of specific industries and sectors, there will be structural similarities (and lessons) across sectors and companies that adopt such technology. This is not only because of the nature of AI technology itself but also because AI cannot be effective in a vacuum. Intelligence is, after all, a marker of human ability and achievements, which implies that widespread adoption of AI has to occur in a socio-technological context. Hence, by necessity, a significant portion of this book will also be concerned with the socio-technological context, including the regulatory landscape and the modern workplace. Implicit in our thesis also is the notion that, because of the nature of the modern inter-connected, innovation-demanding economy, companies will have to eventually adopt any such technology or risk going extinct. Already, there is evidence that winners and losers may be emerging [11].

When we consider the annals of AI history through standard textbooks [27], one of the oldest established areas of AI is *combinatorial search*, wherein a computer program or "algorithm" attempts to solve a problem by searching through a vast space of solutions. Imagine navigating a maze, for example. Assuming the maze is solvable, one could try all possible paths (or all possible "combinations" of steps) in turn, but even intuitively, we can sense that there are more intelligent ways to search for a viable path out of the maze rather than just trying random combinations of steps. The goal of a good AI search strategy is to encode these intuitive insights and "heuristics", sometimes with computational and mathematical guarantees, into a program such that it can strike upon a solution much faster.

Furthermore, although we have used the maze above as an example, it turns out that many real-world problems can be modeled in abstract terms, such that if a good search algorithm can be found for that abstract problem, it could be used to also solve the real-world problem, at least satisfactorily (and in some rare cases, *optimally*, where the machine is able to discover the best possible solution under certain assumptions). In the modern era, applications of such optimization-based search include robot assembly planning, protein design, transportation scheduling, and route-finding for GPS-driven navigation (a common feature on our phones and apps).

These applications did not emerge overnight, nor did the AI that is routinely integrated into them. Following some of the successes and limitations of logic-based AI in the 1960s and 1970s, the AI community as a whole started to wrestle with the uncertainty of the real world. The necessity for more data-driven ("inductive" versus "deductive") and probabilistic approaches started to become more apparent. AI also started to see greater cross-fertilization and collaboration with other fields that could offer new insights, such as information theory, operations research, statistics, probability theory, and control theory. Different areas would eventually become more influential in different subfields of AI, e.g., control theory in robotics, and operations theory in route-planning and search-like problems, which often have a close connection to optimization.

This is not to say that logic, as well as more traditional deductive approaches to automated reasoning, did not continue to be important (and are still important in certain areas and applications of AI). Modern applications include verifying correctness of Very Large-Scale Integration (VLSI[2]) designs prior to production, as well as verifying correctness of cybersecurity protocols and software systems prior to real-world deployment. Logic and deductive reasoning is important in such high-stakes domains because guarantees are often necessary, and stakeholders are uncomfortable with completely trusting a black box to make such decisions (without explanation or proof). Another benefit of using more deductive approaches in such domains is that the problem can very often be modeled as search, and due to decades of research and optimization, efficient algorithms now exist for conducting a near-optimal search (in practice). Other applications include verifying complex bodies of logical rules in applications such as data systems maintenance, processing of insurance claims, security access control, and even tax calculations. Constructed properly, domain-specific reasoning engines can construct *provably correct* plans at scales in such practical areas as manufacturing and logistics. For this reason also, it is not always evident whether an algorithm or problem should be referred to as an "AI" problem or as an operations research problem. The difference between them is more social than fundamental (e.g., although each overlaps significantly with the other in terms of their mathematical and scientific approaches, they still tend to involve different sets of researchers, conferences, publication venues, and even academic departments). In more modern applications, logic and deductive approaches have been applied in tandem with the more inductive approaches we mentioned above. The field of *knowledge graphs* (KGs), which is especially prominent in Web search applications, offers a good case study of such synergy [20, 21].

Notwithstanding the importance of deductive approaches in some domains, since the 1980s, machine learning methods have slowly come to dominate AI, in terms of various objective metrics (such as numbers of publications, researchers, citations, patents, publicized applications, adoption in industry, and so on). Before the deep learning revolution of the last decade, Bayesian networks and probabilistic graphical models, as well as classifiers now considered classic, such as support vector machines (SVMs) and random forests, experienced great popularity from

---

[2] Defined as the process of creating an integrated circuit by combining thousands of transistors into a single chip.

the early 1990s to early 2010s [27]. Probabilistic models have now been used for modeling and diagnosis (as well as monitoring) of complex systems such as jet engines, intensive care patients, and even Mars rovers. Bayesian networks also have a close, and often under-appreciated, connection to causality [22], which has always been an interesting problem in AI, but is taking on new importance today, when many expect AI to explain itself or its decisions (possibly through causal reasoning) in applications of interest. Causal reasoning will also be critical if AI were to be adopted in disciplines that have become very empirical, such as epidemiology and social science.

Coming finally to the last decade, it is hard to argue against the success of deep learning. We could well claim that books such as this one (that are premised on the adoption of AI in organizations being inevitable) would be speculative or practically irrelevant without the documented success of deep learning in problems ranging from object recognition in images, automatic speech recognition and transcription, and automatic inter-translation of human languages (a problem known as "machine translation"). In some exciting domains, such as protein folding, deep learning has achieved success that many assumed were 50 years away [25].

At the same time, the theoretical and practical limits of deep learning are still an active area of study. One argument is that, as we learn more about these networks and how to effectively train them with less data and more energy efficiency, they will prove to be the missing "key" to Artificial General Intelligence (AGI), with a single AI architecture or system capable of doing many of the things that are currently squarely within human capabilities. Others point to some strikingly obvious cases that deep learning gets wrong (e.g., mis-classifying images when minor distortions, invisible to the human eye, are introduced) to argue that it may be fundamentally limited. Regardless of where the future goes, the current generation of deep learning models alone, if adopted properly in various industrial sectors, would likely have lasting and noticeable impacts on productivity and performance.

In the next section, we dive further into the fuzzy relationships between machine learning, AI, and deep learning, including the different types of machine learning. Following that exposition, we turn our attention to Industry 4.0, and how we view it as intersecting with the conception of industries of the future that we adopt in this book.

1.3   AI, Machine Learning, and Deep Learning

One way to think about the relationship between AI, machine learning, and deep learning is that AI is a "super-set" of machine learning, and machine learning is a "super-set" of deep learning. However, it is necessary to qualify the word "super-set." Does it refer to capabilities, i.e., can AI do things that deep learning cannot? This question is a controversial one and has spurred much debate in the AI community. The question becomes somewhat less controversial if we were to consider classes of techniques, rather than capabilities, when discussing AI as a super-set of machine learning or deep learning. For instance, certain techniques developed in the AI planning community would most definitely not be considered machine learning, let alone deep learning. Within the machine learning community, prior to the recent advent of powerful deep neural networks, a number of learning algorithms have been (and in many applications, continue to be) popular, including decision trees, Bayesian networks, and random forests.

With the understanding that all deep learning can be classed as some form of machine learning (although all machine learning is not deep learning), and similarly, machine learning is a *strict subset* of the broader AI community in the sense stated above, we now dive deeper into the three main types of machine learning.

1.3.1 Types of Machine Learning

Generally, machine learning can be divided into three sub-areas:

1. **Supervised Learning:** In this kind of machine learning, a program "learns" from a dataset that has been labeled (usually by humans). The program typically learns a function that can then be applied to a dataset (which has not been seen by the machine before) and predict the labels using the learned function. The performance of such a system is evaluated by comparing its predicted labels to human-annotated labels.

   An example of supervised learning is predicting whether a Twitter account is actually controlled by a human or by a bot. This is a real-world problem with several solutions proposed in the literature, although it is still not completely solved. To "train" the algorithm, we may first provide it with some accounts that we know to be human-controlled and some accounts known to be bot-controlled. In traditional machine learning, we would also extract numeric "features", such as the number of hours the account is active in a day, the number of messages it has published, the number of followers it has, the number of retweets, and so on. Mathematically, the algorithm would now get as input a training dataset with real-valued labeled "vectors" or lists of numbers. The algorithm would learn to predict the label (which we can imagine as 0 for bot-controlled and 1 for human-controlled) using only the vector.

   One can immediately see the utility of both the features and the number of labeled vectors provided to the algorithm as training data. During the "testing" phase, we would now extract the same features from a set of accounts that the algorithm has never seen before and ask it to predict, using its "learned model", whether each account is bot-controlled or human-controlled. One way (albeit not the only way) to quantify how well the model did is to measure its accuracy, which is defined as the fraction of test cases where its prediction agreed with the human label.

2. **Unsupervised Learning:** An obvious limitation of supervised learning is its reliance on labeled data. Many problems do not lend themselves naturally to this framework. In some applications, the goal instead is to discover "structure" in the data. For example, given a social network of individuals, the goal may be to discover "clusters" of individuals who are part of the same community (a problem

that goes by the name "community detection"). Note that we are not seeking to classify individuals, or even pairs of individuals, with labels. Intriguingly, a simple version of this problem can be framed in terms of supervised learning wherein we assume that the input is always a pair of individuals, and the output is 1 (share community) or 0 (do not share community). However, the more powerful version of the problem requires us to discover entire collections of individuals who belong to a community. Computationally or practically, such a problem cannot be addressed by supervised learning. Instead, we have to use an *unsupervised* machine learning algorithm to discover (in this case) the communities in the social network.

Although they both seem to be making predictions, evaluating unsupervised learning can be far more nebulous than evaluating supervised learning. In some applications, the best course of action is not to evaluate it on metrics like accuracy at all, but to instead use it to visualize and explore the data. In machine learning literature, we usually assume that an underlying set of labels is available, annotated by a human. A machine is then considered to have succeeded in unsupervised learning if it ended up grouping items (for example) in such a way that items placed (by the algorithm) in the same group have the same label, and items not placed in the same group have different labels.

3. **Reinforcement Learning:** In contrast with supervised and unsupervised learning, reinforcement learning is inspired by behaviorist principles in psychology, whereby the learning "program" tries out actions in an environment based on such inputs as the current state of the environment and its previous action (or an opponent's previous action when, for instance, playing a game). The agent then receives a reward as a consequence of its actions. The reward may not be immediate, i.e., if a program is playing chess and makes a move, it may not be evident for quite a while (and in some cases, until the end of the game, when the reward is a positive signal to the player who won the game), but it is assumed to occur at some point in time. The program aims to use such rewards to improve its policy, such that it eventually learns a good enough policy to master the game or other decision making problem that it is being trained to do.

Recall from our earlier discussion on AI that efficient exploration of the search space, which tends to be combinatorially explosive, is a fundamental problem that arose in the earliest days of AI. Reinforcement learning has to negotiate such a search using what is commonly called the *exploration–exploitation* tradeoff. In other words, when making a decision, an agent has to ask itself whether it should explore unknown action-spaces and their consequences or go with what it already knows. Overly favoring the latter may lead to a severely suboptimal solution since the agent may be closing itself off to discovering a policy that is far superior to its current best policy.

Similar to both supervised and unsupervised learning, however, over the last decade, deep learning has led to rapid advances in reinforcement learning. The company DeepMind, acquired by Google, is at the forefront of some of this work. They have shown that deep reinforcement learning can be used to achieve superhuman performance even on combinatorially explosive games like Go and, more recently, have achieved breakthroughs in tasks such as automatic protein folding prediction (AlphaFold).

These areas tend to be treated in a disjoint manner in publications, with applications identified as instances of one or the other, but there are many common techniques and architectures (of which neural networks and deep learning are examples) underlying all three. It is easiest to distinguish the three areas in terms of inputs and outputs, rather than through the actual algorithms and techniques employed. For example, deep neural networks have been used for all three kinds of learning noted above, but the inputs to supervised learning are still labeled samples, while even deep reinforcement learning assumes that it will receive reward signals from the environment (at least periodically).

However, even the input–output view can be simplistic, revealing the inadequacy of considering machine learning only along the three dimensions. For example, there has been much work in the machine learning community on *semi-supervised learning*, wherein a much smaller fraction of training data is labeled than would be expected in purely supervised learning. Another example that is fairly well explored in the community is *active learning*. Returning to the previously introduced example on bots, suppose that only a fraction of the Twitter accounts being considered for classification (as human-controlled or bot-controlled) are actually known to be controlled by bots or humans (or to use the terminology introduced, are *labeled*) and the rest are unknown. Of course, we could adopt the purely supervised learning definition of the problem by training a classifier on the (much smaller) training set of samples and then applying it on the (much bigger) test set of unlabeled data. The problem with doing so is that performance will likely be poor if the training set is much too small. Instead, if some amount of manual annotation effort is acceptable, a better option might be to train an *initial* model using the existing training set and then progressively expanding the training set by selecting promising (unlabeled) examples to manually annotate. The premise of active learning is to select samples (for labeling) on which the *current* classifier is most uncertain, since these samples are expected to make the classifier better.

More advanced variants of this kind of "low supervision" machine learning include zero-shot learning, where the *class* of labels in the test set may be from an unknown set. Although the initial training set may be large, the labels within the training set are from one set, but the task definition is such that, in the test set, there may be examples from a class that was never seen during training. For example, a computer vision classifier may have been trained to recognize leopards and lions, but in the test set, images of tigers are also introduced. This is why the problem is called "zero shot," since the system is expected to recognize (at least the presence of) classes that it may not have encountered before, a statistically challenging problem.

1.4   Industry 4.0 Versus Industries of the Future

The phrase "Industries of the Future" is reminiscent of Industry 4.0. We prefer not to use the latter because, in practice, it has come to be associated with specific technologies, not all of which involve AI. These include advanced analytics and predictive models that have been revolutionized by Big Data, Internet of Things (IoT) technology that aims for a revolutionizing of the factory floor (and potentially, homes and offices) by collecting data using Internet-connected devices at the sensory level and optimizing for maximal efficiency and productivity, and blockchain systems that aim to decentralize records of any transaction that can be conducted digitally (which, in the modern age, encompasses almost all transactions, including payments and currency exchange).

At the same time, modern AI paradigms such as reinforcement learning and deep learning have had much impact on typical Industry 4.0 sectors, such as manufacturing and cyber-physical systems. For example, based on a review [23], "smart manufacturing" has now become feasible because of such technologies wherein AI and other adaptive systems can be used to calculate an appropriate dynamic response to complex and variable inputs such as product demand, enabling fast and responsive (even real-time) optimization of inputs and production along the full value chain. A common thread that often emerges from reviews such as these is that AI can be used to make existing systems more flexible, dynamic, self-optimizing, and responsive. In an uncertain world where things can change quickly (both the COVID-19 pandemic and the Ukraine–Russia War serve as useful international examples) and where enterprises are forced to become more nimble to stay productive, such responsiveness may be the key to long-term survival for a company.

To conclude, despite the ubiquity of the term Industry 4.0, its ambiguity and association with certain technologies (including not only AI but also technologies such as IoT), rather than actual industries, makes it constraining to discuss companies and sectors that are traditionally not considered Industry 4.0 (such as AI-based analytics in legal domains that rely on advancements in, for example, natural language processing) but will likely be an important *industry of the future*. This is the terminology that we favor in the remainder of this book. There is already some evidence that the term is also favored by others [29] who feel that it is more appropriate (arguably) than the term Industry 4.0 that already seems to be associated, in practice, with technologies like IoT. In the next section, we discuss one such important piece of evidence, namely, testimony from the Director of the US National Institute of Standards and Technology in 2020 before the US Senate.

1.5   Other (Non-AI) Drivers of Industries of the Future

Although our primary focus in this book is on AI as an enabler for industries of the future, it is useful to consider other drivers of Industry 4.0 as further context. We do not claim that there is universal consensus on the drivers discussed below, but many consider these to be relatively uncontroversial. The drivers are derived from an *Industries of the Future* testimony by Dr. Walter Copan, the under-secretary of Commerce for Standards and Technology, and Director of the US National Institute of Standards and Technology (NIST) in 2020 [29]. The testimony was delivered before the Commerce, Science, and Transportation Committee in the US Senate. Therein, Dr. Copan testified that the "Industries of the Future (specifically Quantum Information Science, Artificial Intelligence, 5G, Advanced Manufacturing, and Biotechnology) were all identified as technological domains that have the potential to transform U.S. manufacturing, communications, health care, transportation, and beyond."

In the rest of this section, we briefly discuss these drivers and also discuss their connection (if any) to AI. We exclude a discussion of AI itself, since the rest of the book is given over to that purpose. Although the drivers are technologies and fields of study in their own right, with no *necessary* connection to AI, in almost all cases, they have the potential to both impact and be impacted by AI in significant ways. Where applicable, therefore, we comment on Dr. Copan's testimony on AI (and use other research as well) to connect the discussion on the other drivers, to AI.

1.5.1   Quantum Information Science (QIS)

In a National Science Foundation (NSF) workshop that was conducted in 1999 [12], well before the current general interest in quantum computing from academia and industry alike, quantum information science (QIS) was described in the report as "an emerging field with the potential to cause revolutionary advances in fields of science and engineering involving computation, communication, precision measurement, and fundamental quantum science."

The goal of QIS is to understand how some of the fundamental laws of physics discovered earlier in the twentieth century can be utilized to radically improve information acquisition, transmission, and processing. Both during the 1990s and even more so since, a growing body of scientists, researchers and technologists have come to see the promise of that goal, which necessarily transcends traditional disciplinary boundaries. More recently, it has become apparent that QIS, and quantum computers, could well become critical to US competitiveness in information technology owing to their potential to change the ways in which we program and compute.

The importance of computing power is central to some of these discussions. Although rapid advances in "miniaturization" of circuitry on silicon chips (guided by Moore's law) have yielded nothing short of a modern revolution in personal and scientific computing, there is an obvious limit to how far it can go. If the trend predicted by Moore's law continues to hold, for example, the 1999 report predicted that miniaturization will reach atomic scale in only two decades. That time has now arrived. If we are to keep realizing constant growths in computing power, a new paradigm will be needed; furthermore, it has become painfully apparent that for many "domain-specific" applications (including AI and machine learning), the ordinary computing paradigm may not be sufficient.

Furthermore, once atomic scale is reached, quantum physics naturally starts playing an important role. As all students of quantum physics know, reality at such small scales (as we understand it) starts to defy "common sense." Observation plays an important role in quantum science, and without observation, it is possible for a particle to behave as if it were in more than one location. Such effects were considered nuisances (at least for the purposes of engineering) only a short time ago, but in QIS, they are treated as a feature rather than a bug. Indeed, one of the QIS's most exciting premises is that such effects can be exploited to perform otherwise-intractable computing tasks. The best example may be the creation of an "unbreakable" code, and another less nefarious example is the design of a working quantum computer. According to some estimates, such a computer would be capable of solving certain problems that would not be solvable even on today's super-computers (unless one was willing to wait out the age of the entire universe and then some).

Because of its potentially revolutionary impact on the speed of computation, QIS could herald an exciting era in AI and its implementation in industries of the future. Of course, realizing QIS in practice in a quantum computer is no trivial matter. It is also noteworthy that, although the "death" of Moore's law has been predicted for many years now, it has continued to hold remarkably robustly, albeit there are questions about whether it is slowing down, and whether the traditional (or non-quantum) computer architecture, no matter how fast, can ever practically be capable of the kinds of complex optimizations that new AI techniques may require for ever larger datasets. Data-intensive machine learning, such as deep learning, has benefited more in the recent past from specialized computing units such as graphics processing units (GPUs) and tensor processing units (TPUs). However, recent progress in quantum computing has been promising. At the time of writing, several mainstream companies seem to be quite far along in their efforts. For example, IBM already has a fleshed-out public-facing website on quantum computing [18], wherein they state that they are "pioneering specialized hardware and building libraries of circuits to empower researchers, developers, and businesses to tap into quantum as a service through the cloud, using their preferred coding language and without having to be quantum experts themselves." The day may not be far when we will be able to tap into quantum resources in the same way that we are able to tap into storage and powerful computing servers (including GPUs and TPUs) through cloud service providers like Amazon, Microsoft, and Google.

1.5.2   5G and Advanced Communication

Wireless communication technology is an emerging technology that has rapidly advanced over the last few decades. There are tens of billions of connected devices globally, far more than the number of people. According to Accenture [30], a substantial fraction of these (hundreds of millions) are, or will be, connected to

5G wireless networks. This will unlock economic value through two channels: modernization of infrastructure and spectrum availability. The adoption of 5G, and infrastructure investments required thereof, could potentially unlock half a trillion dollars of Gross Domestic Product (GDP) in the USA and create more than three million new jobs.

Without going into the nuts and bolts of the technology itself, the most obvious impact of 5G (from an industrial standpoint) will be in IoT. These devices will now be able to collect and send more data, enabling better, more real-time response to changing inputs and conditions. However, even beyond IoT, consumers will benefit greatly from 5G once the infrastructure is fully in place. Multi-modal data generation and consumption will become the new equivalent of text messaging; indeed, the rise of TikTok and success of Instagram highlights first-hand the need for more multi-modal (versus text-based) channels of communication, since these are more expressive and natural to humans.

More consumption and generation of multi-modal data will intersect with AI in a number of ways. First, image and video search will become more commonplace, and as users have ubiquitous access to video, Big Tech companies will feel the need to develop more advanced algorithms (such as unstructured search over video content, not unlike what we are used to today over web documents). Advertising and recommendation systems will also have to adapt to communicate with the user at multi-modal touchpoints. An industry of the future that could rapidly emerge as a consequence of cheaper 5G, and for which there is huge demand, is gaming. Open-world games can be important sources of revenue, as the success of the Grand Theft Auto and Halo series has illustrated, and some games rely heavily on in-game purchases as their revenue model. With 5G, playing such games using cellular connections alone, or over a fast internet connection, will become more feasible and user-friendly. Other tasks that we currently see as mundane may also potentially become "gamified" leading to productivity boosts.

1.5.3   Advanced Manufacturing

As the authors of [2] point out, manufacturing today is entering a sustained period of innovation "driven by the increased integration of sensors and the Internet of things (IoT), increased data availability, and advances in robotics and automaton." There is obviously a role for AI, and especially machine learning, to play here. Some such applications include AI for manufacturing system optimization, manufacturing applications of human–robot collaboration, process monitoring, diagnostics and prognostics, and manufacturing process control.

A full review of these is beyond the scope of this chapter, but an important reason behind the unlocking of such possibilities is the huge amounts and variety of data being collected by smart devices and factories. For example, environmental data can be collected using ambient sensors, allowing accurate recording of room temperature and humidity in real time. Other examples include process data that allows real-time monitoring of heat-treat temperature, energy and machining coolant temperatures; production operation data that can be recorded by controller systems, including machine downtime, shift scheduling, and elapsed time of individual parts in an operation station; and measurement data, such as product diameter and form, from product quality inspections.

Both individually and merged together, these datasets offer opportunities for pattern discovery that may allow organizations to transform large complex man-ufacturing datasets, commonplace in the factories of today, into actionable insights. Such insights could help achieve greater energy efficiency, which could result in many millions of dollars of savings for companies that operate large, innovative factories (such as Tesla), but also predictive maintenance (which would lead to reduced factory downtimes).

1.5.4 Biotechnology

Like many complex fields, biotechnology can be sub-divided into several areas of commercial and research importance, including agricultural biotechnology, animal biotechnology, medical biotechnology, bioinformatics, and industrial biotechnology [31]. Of these, as suggested in the name itself, bioinformatics and computing have had a long history, and more recently, AI has started to play a prominent role in many bioinformatics applications. Bioinformatics is concerned with collection, storage, processing, dissemination, analysis, and interpretation of biological and biochemical information using tools inspired by mathematics and computing to derive insights that can be useful in applications such as biopharma and person-alized medicine. Typically, the information that is collected, stored, and analyzed comprises true Big Data and involves large data pools and databases. Example use cases where AI has a role to play include DNA sequencing from the large datasets, classification of proteins (including its biological function and catalytic role), gene expression analysis, genome annotation, and computer-aided drug design. By way of a specific example of how AI can help add value in biotech applications, Table 1.1 shows some of its primary applications in a prototypical biopharma value chain [31].

Another important use case of AI in biotechnology that has not received nearly as much attention is *agricultural* biotechnology. Agricultural biotechnology is typi-cally characterized by the development of genetically modified plants to increase crop yields, as well as to genetically enhance or improve (or otherwise change select characteristics of) existing plants. While controversial in some respects, it has had the impact of improving agricultural productivity from an economics standpoint. Techniques used in agricultural biotech include both conventional plant breeding and genetic engineering and molecular breeding, micropropagation, and tissue culture. Firms are now looking for AI and machine learning to add value to the agricultural biotech sector by drawing on recent advances in robotics, computer vision, and deep learning. Autonomous robots, for example, can aim to automate at least part of the time-constrained (and seasonal) task of harvesting crops much faster, and potentially with less waste, than is manually possible. It may also reduce dependence on seasonal workers. Data captured by drones deployed in fields

Table 1.1 Some important AI applications in important components of the bio-pharmaceuticals ("biopharma") value chain, based on material covered in [31]

Value chain component	AI application
Research, discovery, and development	Aggregating/synthesizing data; data mining for identifying drug targets and molecular interactions; understanding the mechanisms of disease; and generating and selecting promising drug candidates
Early clinical development	Using AI for more effective trial design, site selection and optimization of participant recruitment; process optimization; proactively predicting toxicity and risk; and monitoring drug adherence and procedure compliance
Late clinical development	Supply chain planning and optimization; predictive maintenance; inventory management; demand forecasting; logistics and optimization; workforce planning; medication and tracking
Commercialization, marketing, and launch	Patient engagement; physician and clinical decision support; predictive pricing; effective and optimized marketing/advertising; and launch coordination
Post-market surveillance, patient support, and follow-up	Tracking adherence to medication; recognition of adverse events; monitoring compliance; and designing optimal patent and clinical support programs/platforms

can also be processed and analyzed by computer vision algorithms, which would help to monitor soil health and crops in real time and help predict and preempt problems before they arise. Other machine learning applications include modeling and predicting environmental and weather changes that have direct impact on crop yields.

1.6   Where Will Industries of the Future Come From?

The question of *what* specific industries of the future will be is at least as challenging to forecast about than the question of *where* they will come from. Indeed, innovation is fundamentally hard to predict, and AI is particularly susceptible to faults in both over-prediction and under-prediction. Undoubtedly, however, the twin questions of where and what are intertwined here. While the innovations may occur in research (the role of which we subsequently discuss) in both industry and academia, the actual commercialization of these innovations will require industry presence. In Chap. 3, we consider the question of where AI-based industries in the future will emerge from in a lot more depth, including by drawing on recent and specific examples. Herein, we preview briefly the three likely sources from which such innovation will emerge in a commercialized form and become mainstream over time:

1. First, in the technology space, startups have always been a prominent venue for commercialization of innovative products and services that end up becoming market creators in their own right. In the most recent boom, startups valued at over a billion dollars (called "unicorns") were not uncommon. Many such startups rely on AI at least in part. Self-driving vehicle automation was a particularly hot source of funding the previous decade, but other sectors that have been beneficiaries of the capital with which venture capitalists and private equity were flushed include educational technology (EdTech), life sciences, natural language processing, and productivity-enhancing software. AI, as understood, plays an applied role in some of these companies. However, some companies are centered on AI and have prominent backers. A good example is OpenAI, which describes itself as an "AI research and deployment" company. Another example is Hugging Face, which hosts many language models (which have arguably revolutionized natural language processing in the last few years) and that raised its most recent round of funding of $100M (at the time of writing) in May 2022 [5]. The pace and density of fundamental AI research is high in these companies, suggesting that (whether as acquired, or eventually, publicly listed, companies) they may play a keystone role in industries of the future that are premised on AI advances.

2. Second, the current generation of Big Tech companies (and their Chinese counterparts) has been aggressive in its acquisition of innovative companies. The Big Tech firms are also active in research, with many releasing open-source packages and models that have been enthusiastically adopted by the AI community (in academia and industry alike). Researchers in these firms tend to publish in the top AI and machine learning conferences, and many graduates today view a full-time position in these research labs to be as prestigious, if not more, as an academic position in a major university. Clearly, these strategies are currently bearing fruit. Indeed, some of the most noteworthy initiatives in AI that have eventually been commercialized through these companies have almost all had their origins as acquired startups. Examples include (but are not limited to) the Siri voice assistant that has become a staple of the Apple ecosystem, the Google Knowledge Graph (which is not as well known to the general public but plays an important role in the Google search engine today), and the stunning AI-driven scientific successes of DeepMind (also acquired by Google) in areas ranging from bioinformatics to gameplaying and even rudimentary Artificial General Intelligence, to only name a few. Whether this innovation will continue (especially as economic headwinds appear) or will be the first casualty once the tech sector as a whole is compelled to re-orient, due to factors such as plummeting stock prices (or regulation), is anyone's guess. Taken as a whole, it seems likely that the sector will play an important role in the *long run* in developing industries of the future. We note also that, even though these companies are viewed as "tech" companies, they have not been shy about moving into markets not traditionally considered as Big Tech territory. For example, rumors of a self-driving car being developed at Apple (allegedly code-named Project Titan) have

now been circulating for a while [16], and some of Google's subsidiaries have vested interests in healthcare and life sciences [4].

3. Third, AI has now matured to the point where non-tech companies, including large corporations in sectors such as auto, healthcare, and even real estate development, construction, and sales, are exploring its uses. For example, AI is able to modify images of rooms in houses with "virtual" furniture, a process that is called "virtual staging" and is being used by ordinary people to make pictures of their houses more appealing on platforms like Zillow (where prospective buyers of real estate often browse, at least during their initial search). The images look photo-realistic, but their generation is non-trivial. This is a good example of how existing state-of-the-art AI methods and algorithms, appropriately tuned for a domain-specific application, can be used to improve real-world outcomes. Although it is unlikely that fundamental AI innovation will take place within these companies, the influence of large non-tech corporations (as funders, customers and adopters) of such innovation cannot be over-stated. It should be noted that many of these corporations have venture arms that are often on the lookout for such innovation to acquire or incubate, especially those that could have a very real impact on their future business. Such companies also tend to take a more pragmatic view of how AI will impact their bottom line and can be an inspiring source of ideas and needs for nascent startups.

1.7   The Role of Research

In a knowledge-intensive economy, the role of research is increasingly important in the development of risky and innovative technologies. AI is no different and may be even more innovation-sensitive.

According to Neo-Schumpeterian ("new-growth") economics, innovation, knowledge, and entrepreneurship are the three major forces driving economic dynamics [17]. Since novelty is a key distinguishing mark of Neo-Schumpeterian economics, innovation (as the most visible manifestation of novelty) is probably the most important of these sources. Knowledge (and especially, scientific knowledge) complements technological innovation in a dynamic environment, especially with rapid growth of data and computation. Within this framework, entrepreneurship is the end result for realizing gains accrued from advanced innovation that has commercial potential. Within AI, and technology more broadly, startups have always been an important source for such commercialization, serving as market makers for a service (such as Uber or DoorDash) where none existed before.

Much of the research underlying modern technological innovations has historically been conducted in an academic lab as *fundamental research*. Some of this research then gets applied to real-world problems and incubated in startups. While this pipeline is still largely intact (albeit at a much more rapid pace than it historically took innovations to be commercialized[3]), it must be noted that AI fundamental research is not the exclusive province of academia in our current day and age. Companies are aware of the economic driving forces, with all of the Big Tech firms placing a premium on innovation, and consequently investing heavily in R&D. As we discuss in later chapters, many of the firms have substantive and well-staffed research labs; in the case of subsidiaries like DeepMind, one might argue research is the primary and most valuable output being produced. Researchers in Big Tech are also fairly well established in their fields, with a good record of publications and citations in their respective areas. A sizeable number of patents are generated by these firms,[4] but so are papers in the top AI conferences and journals. Tables 1.2 and 1.3 provide concrete recent examples of the kinds of AI and other emerging tech innovations being patented by four Big Tech firms. Table 1.4 provides examples of papers published in 2021 by researchers in these companies in a top peer-reviewed machine learning conference .

Certain research institutes like the Allen Institute for Artificial Intelligence (AI2) were not incubated in a university, although they harbor close connections with academics in universities such as the University of Washington at Seattle. These are important sources of original research and innovation. Founded in 2014 by Paul Allen (one of the co-founders of Microsoft), AI2 is a non-profit research institute that describes its mission as "conducting high-impact AI research and engineering in service of the common good" [1]. The institute is quite diverse in its portfolio of AI projects undertaken, which (among environment-related and climate modeling projects) include EarthRanger (protecting wildlife with data), Skylight (improving maritime transparency), and Climate Modeling. There are also ambitious natural language processing projects on commonsense and AI and teaching AI to read and reason. More recently, the institute has also launched an incubator for AI-related startups.

The role of research institutes like AI2 and research labs in Big Tech notwithstanding, academia remains (in our view) the most important and prolific source of AI research. This is evidenced not just by the volume of publications (most of which are heavily empirical, and many of which look to real-world use cases and applications for inspiration) in the important conferences and journals, but also the volume of R&D funding that academic labs benefit from. The federal government, at least within the United States, is an important source of this funding. While funding for fundamental scientific research in the USA is often attributed to the National Science Foundation (NSF), agencies like the Department of Defense , especially through the Defense Advanced Research Projects Agency (DARPA), are also well-known funders of fundamental, moonshot AI research in academia and

---

[3] As case in point, technologies like the smartphone and the touchscreen were invented decades before they were commercialized into products that are now ubiquitous.

[4] The exact numbers of these patents depend on a thornier question of which patents count as AI patents. However, the press has been covering the issue of ramped-up patent activity by Big Tech in some detail [32].

industry alike. Table 1.5 notes some recent DARPA programs and their objectives
that have specifically sought to push the current frontier of AI through high-risk,
high-reward research. Even the NSF recognizes the importance of commercializing
viable research products via their Small Business Innovation Research (SBIR) and
Small Business Technology Transfer (STTR) programs, widely known as the NSF's
"seed fund." Per their website [13], the focus is on "transforming scientific and
engineering discoveries into products and services with commercial and societal
impact." They have already awarded $200+ million in R&D funding to more than
400 startups across the United States.

These trends in funding are likely to intensify as the US federal government
has explicitly recognized in recent years that, without ramping up investments in

Table 1.2 Examples of patents granted in the United States in 2022 to Facebook (Meta) and
Amazon. Note that the patents were filed earlier than 2022

Patent number	Patent title	Company	Abstract
11373640	Intelligent device grouping	Amazon	Systems and methods for intelligent device grouping are disclosed. An environment, such as a home, may have a number of voice-enabled devices and accessory devices that may be controlled by the voice-enabled devices. One or more models, such as linguistics model(s) and/or device affinity models, may be utilized to determine which accessory devices are candidates for inclusion in a device group, and a recommendation for grouping the devices may be provided. Device group naming recommendations may also be generated and may be sent to users
11366971	Speech recognition accuracy with natural language understanding based meta-speech systems for assistant systems	Facebook	In one embodiment, a method includes receiving, from a client system associated with a first user, a first audio input. The method includes generating multiple transcriptions corresponding to the first audio input based on multiple automatic speech recognition (ASR) engines. Each ASR engine is associated with a respective domain out of multiple domains. The method includes determining, for each transcription, a combination of one or more intents and one or more slots to be associated with the transcription. The method includes selecting, by a meta-speech engine, one or more combinations of intents and slots from the multiple combinations to be associated with the first user input. The method includes generating a response to the first audio input based on the selected combinations and sending, to the client system, instructions for presenting the response to the first audio input

Table 1.3 Examples of patents granted in the United States in 2022 to Google (Alphabet) and Microsoft. Note that the patents were filed earlier than 2022

Patent number	Patent title	Company	Abstract
20220207264	Subject identification based on iterated feature representation	Microsoft	A computer vision method includes recognizing a feature representation of a query image depicting an unknown subject. A similarity score is computed between the representation of the query image and feature representations of a plurality of gallery images collectively depicting two or more different subjects with at least two or more gallery images for each subject, and each gallery image having a label identifying which of the subjects is depicted. One or more updated feature representations of the query image are sequentially iterated based on one or more of the computed similarity scores. For each of the one or more updated feature representations, an updated similarity score is computed between the updated feature representation and the feature representations of each of the gallery images. The unknown subject is identified based on a gallery image having a highest updated similarity score
11373049	Cross-lingual classification using multilingual neural machine translation	Google	Training and/or using a multilingual classification neural network model to perform a natural language processing classification task, where the model reuses an encoder portion of a multilingual neural machine translation model. In a variety of implementations, a client device can generate a natural language data stream from a spoken input from a user. The natural language data stream can be applied as input to an encoder portion of the multilingual classification model. The output generated by the encoder portion can be applied as input to a classifier portion of the multilingual classification model. The classifier portion can generate a predicted classification label of the natural language data stream. In many implementations, an output can be generated based on the predicted classification label, and a client device can present the output

emerging and advanced technologies,[5] it will become increasingly difficult for the USA to maintain its present, but diminishing, lead over a geopolitical rival like China that is also investing heavily in these technologies.

---

[5] This includes not just AI but also many of the emerging technologies we discussed earlier in the chapter, including QIS and biotech.

Table 1.4 An example of a 2021 peer-reviewed publication by researchers in one of the four Big Tech companies in the NeurIPS conference, widely regarded as the best machine learning conference

Full citation of paper	Company
Fatemi, M., Killian, T.W., Subramanian, J., & Ghassemi, M. (2021). Medical Dead-ends and Learning to Identify High-risk States and Treatments. Advances in Neural Information Processing Systems, 34, 4856–4870	Microsoft
Fifty, C., Amid, E., Zhao, Z., Yu, T., Anil, R., & Finn, C. (2021). Efficiently identifying task groupings for multi-task learning. Advances in Neural Information Processing Systems, 34, 27503–27516	Google
Hu, Z., & Li, L.E. (2021). A causal lens for controllable text generation. Advances in Neural Information Processing Systems, 34, 24941–24955	Amazon
Tarbouriech, J., Pirotta, M., Valko, M., & Lazaric, A. (2021). A provably efficient sample collection strategy for reinforcement learning. Advances in Neural Information Processing Systems, 34, 7611–7624	Facebook

1.8 Future Developments

Modern AI is a fast-moving field with deep synergies between academia and industry. Leaders of many AI labs are frequently recruited from academia, and exchanges and consulting (or visiting) arrangements between professors in major universities, and the Big Tech companies, are fairly common.

What does this mean for industry? An immediate prediction, if such trends continue to hold, is that given the current state of AI technology, and with moderate advances in dexterity and perception, smart actuators and robots could soon be moving into unstructured environments and performing tasks that have some level of ambiguity. Such environments encompass domains such as warehouses, roads and infrastructure, agriculture, mining, and, worryingly, the battlefield. Language understanding is now starting to equal image understanding in terms of accuracy, domain adaptability, and ease of use, and the applications have already started to emerge. Intelligent personal assistants will only become more intelligent over time, and in fields like EdTech, intelligent tutoring systems (including in virtual reality) may soon become commonplace. Search engines are improving with each year and are able to now serve relatively complex requests, where once (not too long ago) keyword requests with more than 5–6 keywords were barely manageable. It is also important to note that these advances are not occurring in a vacuum. With the concurrent advancement of other sectors of Industry 4.0, several of which were covered earlier, new industries and possibilities will emerge.

One of the consequences of the synergy between academia and industry is that the commercialization of fundamental research in academia is no longer a linear process. Datasets and baseline systems (such as pre-trained language models for natural language understanding) for furthering such innovation tend to come from

Table 1.5 Examples of five DARPA programs, websites and stated purpose (on the website) that have sought to push the frontier of AI through high-risk, high-reward research

Program name/citation	Stated purpose (from website)
Memex [9]	Invent better methods for interacting with and sharing information, so users can quickly and thoroughly organize and search subsets of information relevant to their individual interests ... technologies developed in the program would provide the mechanisms for improved content discovery, information extraction, information retrieval, user collaboration, and other key search functions
Low-Resource Languages for Emergent Incidents (LORELEI) [7]	Dramatically advance the state of computational linguistics and human language technology to enable rapid, low-cost development of capabilities for low-resource languages ... technologies resulting from LORELEI research will be capable of supporting situational awareness based on low-resource foreign language sources within an extremely short time frame—starting as soon as 24 h after a new language requirement emerges
Learning with Less Labeling (LwLL) [6]	Make the process of training machine learning models more efficient by reducing the amount of labeled data required to build a model by six or more orders of magnitude and by reducing the amount of data needed to adapt models to new environments to tens to hundreds of labeled examples
Science of Artificial Intelligence and Learning for Open-world Novelty (SAIL-ON) [10]	Research and develop the underlying scientific principles, general engineering techniques, and algorithms needed to create AI systems that act appropriately and effectively in novel situations that occur in open worlds
Machine Common Sense (MCS) [8]	Address the challenge of machine common sense by pursuing two broad strategies. Both envision machine common sense as a computational service or as machine commonsense services. The first strategy aims to create a service that learns from experience, like a child, to construct computational models that mimic the core domains of child cognition for objects (intuitive physics), agents (intentional actors), and places (spatial navigation). The second strategy seeks to develop a service that learns from reading the Web, like a research librarian, to construct a commonsense knowledge repository capable of answering natural language and image-based questions about commonsense phenomena

industry anyways, and due to the culture of conference-based dissemination of scholarly work in computer science, research tends to diffuse rapidly into industrial practices (especially, Big Tech). All of this implies that fundamental advances in AI will tend to percolate rapidly into industrial products and services, especially once the initial learning curve has been overcome (in non-tech sectors) and some kind of baseline AI infrastructure is in place within the organization.

In the long-term, a critical goal (since the very inception of AI) has been general-purpose AI or Artificial General Intelligence (AGI). The jury is still out on when AGI will be achieved, but if and when it is achieved, it may become a substitute

for labor. This will likely have massive economic ramifications, although some of
it is expected to be beneficial (such as providing assistive care to a rapidly aging
population across both the Western world and countries like Japan and China). In
Chap. 4, we discuss the more immediate workforce consequences of augmented
AI and why human resource managers and C-suite executives need to start paying
attention to this issue today.

At the present time, however, we still seem to be quite far from developing
a truly general-purpose AI. Some have argued that data and computing power
alone will eventually reach their limits and that conceptual breakthroughs in AI are
needed for AGI to ever become a possibility. Technical challenges include building
robust causal reasoning into AI, decision making over long time horizons, and more
intelligent use of knowledge (as well as efficient and selective utilization of new
knowledge).

John McCarthy, one of the founding fathers of AI, was reportedly quoted [26] as
saying in an interview in the mid-1970s that "What you want is 1.7 Einstein's and
0.3 of the Manhattan Project, and you want the Einstein's first. I believe it'll take
five to 500 years." Despite radical progress, including a deep learning revolution
of sorts, the quote holds as true today as it did back then. Nevertheless, because
of these advances and other emerging technologies such as quantum computing,
researchers believe that AGI could be achieved within this century, although there
is considerable variation in their predictions on *when* precisely in this century (they
believe) it will be achieved.

References

1. For AI, A.I.: Allen institute for AI: About (2022). URL https://allenai.org/about
2. Arinez, J.F., Chang, Q., Gao, R.X., Xu, C., Zhang, J.: Artificial intelligence in advanced manufacturing: Current status and future outlook. Journal of Manufacturing Science and Engineering **142**(11) (2020)
3. Ashford, D.: The mechanical Turk: Enduring misapprehensions concerning artificial intelligence. The Cambridge Quarterly **46**(2), 119–139 (2017)
4. Calico: Calico labs (2022). URL https://www.calicolabs.com
5. Crunchbase: Hugging face - funding, financials, valuation & investors (2022). URL https://www.crunchbase.com/organization/hugging-face/company_financials
6. DARPA: Learning with less labeling (LwLL). URL https://www.darpa.mil/program/learning-with-less-labeling
7. DARPA: Low resource languages for emergent incidents (LORELEI). URL https://www.darpa.mil/program/low-resource-languages-for-emergent-incidents
8. DARPA: Machine common sense. URL https://www.darpa.mil/program/machine-common-sense
9. DARPA: Memex (archived). URL https://www.darpa.mil/program/memex
10. DARPA: Science of artificial intelligence and learning for open-world novelty (sail-on). URL https://www.darpa.mil/program/science-of-artificial-intelligence-and-learning-for-open-world-novelty
11. Eide, N.: It's early days, but AI will define industry "winners and losers," KPMG says. (2012). URL https://www.ciodive.com/news/its-early-days-but-ai-will-define-industry-winners-and-losers-kpmg-say/562553/
12. Foundation, N.S.: Quantum information science (1999). URL https://www.nsf.gov/pubs/2000/nsf00101/nsf00101.htm

13. Foundation, N.S.: About America's seed fund powered by NSF (2022). URL https://seedfund. nsf.gov/about/
14. Frank, A.: Are we in an AI summer or AI winter? (2021). URL https://bigthink.com/13-8/ are-we-in-an-ai-summer-or-ai-winter/
15. Gonsalves, T.: The summers and winters of artificial intelligence. In: Advanced methodologies and technologies in artificial intelligence, computer simulation, and human-computer interaction, pp. 168–179. IGI Global (2019)
16. Gurman, M.: Apple accelerates work on car project, aiming for fully autonomous vehicle (2021). URL https://www.bloomberg.com/news/articles/2021-11-18/apple-accelerates-work-on-car-aims-for-fully-autonomous-vehicle
17. Hanusch, H., Pyka, A.: Principles of Neo-Schumpeterian economics. Cambridge Journal of Economics **31**(2), 275–289 (2007)
18. IBM: Quantum computing (2022). URL https://research.ibm.com/quantum-computing
19. Ireland, C.: Alan Turing at 100 (2012). URL https://news.harvard.edu/gazette/story/2012/09/ alan-turing-at-100/
20. Kejriwal, M.: Domain-specific knowledge graph construction. Springer (2019)
21. Kejriwal, M., Knoblock, C.A., Szekely, P.: Knowledge graphs: Fundamentals, techniques, and applications. MIT Press (2021)
22. Pearl, J.: Graphical models for probabilistic and causal reasoning. Quantified representation of uncertainty and imprecision pp. 367–389 (1998)
23. Peres, R.S., Jia, X., Lee, J., Sun, K., Colombo, A.W., Barata, J.: Industrial artificial intelligence in industry 4.0-systematic review, challenges and outlook. IEEE Access **8**, 220121–220139 (2020)
24. Rosenblatt, F.: The perceptron: a probabilistic model for information storage and organization in the brain. Psychological review **65**(6), 386 (1958)
25. Ruff, K.M., Pappu, R.V.: Alphafold and implications for intrinsically disordered proteins. Journal of Molecular Biology **433**(20), 167208 (2021)
26. Russell, S.: The history and future of ai. Oxford Review of Economic Policy **37**(3), 509–520 (2021)
27. Russell, S.J.: Artificial intelligence a modern approach. Pearson Education, Inc. (2010)
28. SHRM: Thousands of British workers begin trial of 4-day workweek. (2022). URL https://www.shrm.org/resourcesandtools/hr-topics/benefits/pages/thousands-of-british-workers-begin-trial-of-four-day-work-week.aspx
29. of Standards, N.I., Technology: Industries of the future (2020). URL https://www.nist.gov/ speech-testimony/industries-future
30. Strategy, A.: Accelerating future economic value from the wireless industry. (2018)
31. Team, Q.: Ai in biotechnology exploring role and applications (2020). URL https://qualetics. com/role-applications-of-ai-in-biotechnology/
32. Watkins, N.: Inside big techs race to patent everything (2022). URL https://www.wired.com/ story/big-tech-patent-intellectual-property/
33. Weizenbaum, J.: Elizaa computer program for the study of natural language communication between man and machine. Commun. ACM **9**(1), 3645 (1966). DOI URL https://doi.org/10. 1145/365153.365168

AI in Practice and Implementation: Issues and Costs

2.1 Introduction

The previous chapter introduced both AI and the concept of "Industries of the Future". Before discussing such industries, and their incubation in both large and small enterprises in Chap. 3, it is important to understand AI's adoption in enterprise today . On the one hand, AI has proven to be an exciting draw for investors and decision-makers alike, with COVID-19 proving to be an accelerant. According to a Harvard Business Review article in 2021 [33], a PricewaterhouseCoopers (PwC) study found that 52% of companies intensified their AI adoption plans as a result of the pandemic. AI was also seen as critical to addressing the skills shortage, which is only likely to intensify in the wake of systemic trends like the Great Resignation.

On the other hand, AI adoption in enterprise has always had a "hype problem." In their *Eye on A.I.* newsletter on March 29, 2022, Fortune writes about how, instead of "creating a market worth trillions," the current market for AI is still in the range of "hundreds of billions of dollars." Although significant, it begs the question of why a market worth trillions of dollars has not arisen yet. Is it, for example, money left on the table (to use a maxim) in that the market for current AI technology *could* be worth trillions of dollars but has just not been realized due to short-sightedness (or lack of enthusiasm in the C-Suite for such investments) on the part of companies? Or is it that the technology itself has not lived up to its hype and that further advancements are needed before such a market will be created, and implementation will take off? Will we have to suffer through another (or more than one) AI winter before the potential of the technology is realized [12, 38]?

Although there is no doubt that more advancements are needed (and in particular, for AI to become more robust, explainable, and "general"), there is also an argument to be made that even current AI technology has many more avenues for providing value than has been explored within industry thus far. In part, cost is always an issue, given the rather abstract nature and value proposition of AI algorithms and advancements in fields such as Natural Language Processing. There is also concern

© The Author(s), under exclusive license to Springer Nature Switzerland AG 2023

M. Kejriwal, *Artificial Intelligence for Industries of the Future*, Future of Business and Finance, https://doi.org/10.1007/978-3-031-19039-1_2

(and indeed, rationally so) about privacy and compliance issues [13], especially in the wake of laws such as the European Union's General Data Protection Regulation (GDPR).

In this chapter, we consider all of these issues and costs that are pertinent to implementing AI in modern-day industry. We start with some major challenges that currently stand in the way of large-scale adoption or that stymie the investments needed to fully weave AI into the fabric of an enterprise. We then consider the important issue of cost. Unfortunately, calculating a return on investment or ROI for AI has proven to be a difficult task, although progress is being made. As one example, we describe a framework that has been proposed in the consulting literature for conceptualizing ROI for AI projects.

2.2   Challenges in Implementing AI

Given the transformative nature of AI technology, a natural question to ask is why it has not become ubiquitous yet. One might cite time and money as challenges, but as economists have shown, over the last decade, companies have tended to use their excess capital and profits for stock buybacks and other "financial investments" (such as Bitcoin reserves), rather than plowing the profits back into capital investments. We have also just witnessed an era of historically low interest rates and inflation, both of which are now rapidly trending upward (at least at the time of writing). In short, companies had the money, yet chose not to engage in a massive investment in AI.

At the same time, it would not be fair to underplay the impact that AI has been having on various industries. (There would be no point in writing a book such as this one if there had been no interest in, or implementations of, AI in industry.) A number of consulting and professional services firms, including Deloitte and Accenture, have been commenting actively in their reports on how AI is becoming a priority in executives' decision making, although many questions still remain on its value and return on investment (ROI).

Our argument for why transformative AI has not become ubiquitous is simpler. Implementing AI may have become easier over time, due to the rise of open-source software packages (including for neural network implementations, such as PyTorch and TensorFlow) and datasets, but still involve a host of engineering challenges and executive buy-in. Modern AI is famously data-intensive, and acquiring, processing, and integrating data to derive value has always been a hard problem, involving specialized highly-paid personnel and careful treatment of proprietary datasets (especially if privacy is a legal or business concern) and internal procedures. The advent of "data science" with a Harvard Business Review article calling data the new "oil" is the perfect analog. Over the previous decade, that view has been gradually dialed back, and rightfully so [41].

AI, although sometimes conflated with data science, is both narrower in some respects (e.g., only cleaning data to make it more suitable for processing is considered a precursor to, but not necessarily a part of, AI and machine learning algorithms unless the cleaning itself involves some kind of intelligent automation) and broader in others (e.g., in application of advanced statistical techniques, which is not usually considered essential in day-to-day data science work).

Beyond the challenges associated with data acquisition and quality, the increased scrutiny of government and regulatory bodies, especially on the practices of Big Tech, companies also have to actively manage privacy concerns and comply with regulations, where applicable. Finally, within the business itself, there may be skepticism among executives in investing a large sum of money in AI if the ROI is not clear. As we subsequently discuss, there are both social and technical problems in deriving a bridge between AI quality metrics on the one hand [40] and financial metrics such as earnings and cash flow, on the other. An understanding of these challenges, which are meant to be representative, not exhaustive, is going to be critical for companies that see themselves as vanguards of future industry and practices and are looking to be first movers in their industries with respect to this transformative technology.

2.2.1 Data Acquisition

Data acquisition is an obvious first step before AI applications can be enabled at scale. In the research community, data acquisition is generally taken for granted, due to prevalence and standardized formatting of acceptable benchmarks, such as ImageNet in the computer vision community [5]. In industry, data can play a key role in cementing the competitive advantage of the organization. An organization may, in fact, choose to use open datasets, or even start exclusively with open datasets, when developing its AI, but at some point, will want to embed its proprietary datasets into the pipeline to derive competitive advantage. In some domains, selling valuable datasets is a business in itself [39], which can be controversial when it comes to issues of (for example) personal autonomy and healthcare. A famous and relatively non-controversial example that is often cited in finance is when hedge funds and quantitative finance specialists acquire satellite imagery [14], or even social media data, to derive signals that may allow them to incorporate information into their models that allow them to make better, more precise predictions of market conditions. Although costly, acquiring data in this manner does have some advantages, since companies selling them have an incentive to make them high quality.

In other cases, the company has to acquire its own data and use it in AI applications. Industry 4.0 applications [36], such as smart manufacturing [31], are good examples, with raw streams of data acquired from sensors and other devices. The timeliness and real-time nature of data acquisition can make a significant difference. Within the Internet of Things (IoT) and cyber-physical systems domains, acquiring data in real time from the "physical" realm in which the sensors are deployed, along with ensuring that the feedback cycle between the virtual and physical realms remains real time, can both impact the decision on whether the system should be deployed in the first place. Data acquisition in IoT involves not only the gathering of information but also conversion of raw information to the data structures that are manipulable in software and by AI, the extraction of relevant knowledge from the data, the exchange of knowledge between sensory modules (if necessary), and the high-level analytics that can then be used in a feedback cycle to configure and optimize the external environment.

Note that *real time* does not necessarily imply *instantaneous*, but a lag of even minutes may be unacceptable for some applications. This can determine design choices such as whether to send data to the cloud for processing or to design more powerful edge computing processors that can do rapid processing on the spot but may be limited in their capabilities and may have power requirements that are infeasible. Given these challenges, the AI may take a backseat to initial deployment and simple analytics, at least initially.

In healthcare, many AI algorithms are designed to take as input electronic health records (EHRs) that are critical for understanding patient health [11]. Although we commented earlier that we will not try to predict, to a great extent, *what* an industry of the future might be, one safe prediction is AI-assisted healthcare. Certainly, more advanced analytics on EHRs that are rapidly becoming feasible and cost-effective due to investments in AI suggest that greater insights will eventually be drawn from such data that document not only operational needs but also actual healthcare delivery. At the present time, however, EHR datasets exhibit considerable variety and irregularity and have traditionally been difficult to input directly to algorithms. EHRs are primarily collected during a patient's treatment to improve the patient's health and also facilitate clinical research [4]. Data acquisition itself is heterogeneous: it may be collected from a broad variety of sources, such as (high-frequency) signals sampled once every thousandth of a second, vital signs that are noted hourly, irregular laboratory (and/or imaging) tests that are only recorded on an as-needed basis, notes written by doctors, specialists or other human stakeholders at transitions of care, and demographic data that remains largely unchanged and only needs be collected once.

Clinical data also tends to be acquired longitudinally [1], whether for a single patient or at the level of the entire hospital. A hospital may, for example, be collecting longitudinal records for all patients with a specific disease. Each such set of records is also heterogeneous, along the dimensions of datatype, timescale, reasons for collection, and so on. Ultimately, using such multi-granular, heterogeneous data to facilitate AI-enabled applications at scale can be very challenging for healthcare providers.

When data acquisition is high-volume, and distinctions have to be made between longitudinal and cross-sectional acquisition of samples, attention also has to be paid to the statistics of the data. Longitudinal data may have autocorrelation [3], violating assumptions of independent and identical distribution (i.i.d.) that are common in many machine learning applications. From an ontological or semantic standpoint, care also has to be taken to carefully distinguish between the data and metadata. Missing data may also be a common problem, as might be obsolescence of data acquired over time. The boundary between data acquisition and data quality (see next section) becomes a gray zone when these concerns arise. Finally, in the case of supervised systems in particular, collecting raw data alone is not enough: the organization must also have an efficient system for acquiring *labels* for the data that can be used to train deep neural networks and other supervised machine learning models. Some companies rely on internal crowdsourcing efforts [45] or teams of "editors" to provide labels, while others may rely on a commercial crowdsourcing service such as Amazon Mechanical Turk. Yet others may choose to contract with, or use the platforms or software of, commercial data labeling [6] services, such as Appen and Dataloop.

2.2.2 Data Quality

In an era of sensors, smartphones, crowdsourcing, and businesses that collect, aggregate, and sell data as a service, the quality of data becomes a paramount issue. Even non-technical stakeholders today are familiar with the term *Big Data*. Intuitively, Big Data refers to datasets are high-*volume* and that cannot "fit on your phone," but in practice, other dimensions such as *variety, velocity*, and *veracity* (which, along with volume, constitute the traditional 4V's of Big Data) are also important. In large organizations, there is a growing recognition that even data that was previously not considered to be useful (such as logs of uncritical systems) could, in fact, be leveraged to yield insights at low cost.

Such logs (and other similar data), which tend to have some structure but are not the kind of data stored in a relational database management system (RDBMS) such as Oracle or SQL Server or have high priority associated with them, are more likely to fall under the practical definition of Big Data. As noted above, veracity is one of the dimensions of Big Data and the one that correlates with *data quality*. Veracity implies that it may not always be possible to trust Big Data or that considerable quality control may be needed (such as deduplication[1] [18], schema matching [42], and removal of elements that may be over-represented in the data but do not serve a practical purpose). However, quality issues tend to arise even in "primary" data, such as the financial or personal records in an RDBMS. Unfortunately, research on data quality is still very much in its infancy compared to aspects such as volume or velocity (for which infrastructure, including for storing and analyzing streaming data) that were the initial frontiers for value extraction in companies.

Healthcare is an example of an industrial sector where data quality is of extreme importance, since it could influence medical decision making and reporting. Many authors have previously studied issues of data governance in the healthcare sector

---

[1] Scalable deduplication is an even harder problem where considerable research has been invested over the last 25 years. More recent work from our own work include [15, 16, 19, 22, 24], and [20], to only list a few. Other problems that have come under much recent study include *minimally supervised* deduplication, which involves doing deduplication with little to no training data [23,25], and deduplication on highly heterogeneous datasets that may not even have an explicitly defined schema [26, 27]. Experimentally, one must also be sure that the methods will generalize and are not subject to overfitting and other sources of bias [21].

using (what they refer to as) *data quality dimensions* (DQDs) [37]. Five selected dimensions, appropriately named, include:

1. **Completeness:** This dimension measures the extent to which the data is complete enough (i.e., has both sufficient breadth and depth) for the tasks for which it is being employed.
2. **Interpretability:** This dimension measures whether the definitions of data headers and units are clear and whether appropriate languages, units and symbols are being used to describe the data.
3. **Relevance:** This dimension measures whether the data is applicable for the task at hand or if its utility is tangential.
4. **Timeliness:** This dimension measures whether the data is sufficiently up to date for the applications at hand.
5. **Cleanliness:** This dimension measures whether the data is reliable, correct, and largely free of errors.

Although it is intuitively clear why the listed DQDs are expected to be important in healthcare applications, let us take a specific example (personalized decision support systems or PDSS) [34]. A PDSS system aims to enhance personalized medicine in an evidence-based manner through Big Data analytics. One can immediately see the case for having relevant, timely, clean, complete, and interpretable data when implementing or relying on the decisions of a PDSS. Other healthcare applications where such dimensions matter, albeit not to the same extent, are prevention of healthcare (especially, Medicare) fraud, facilitating medical research (especially involving integration of heterogeneous datasets), and quality control measures, such as reducing patients' readmission rates. For this last application, relevance might be a more important DQD compared to (for example) timeliness. In all cases, we hypothesize that interpretability is important, but in applications that are high-stakes or have legal ramifications, it may become paramount. Timeliness, on the other hand, may not always be important in some applications but can become paramount in others (automatically detecting strokes or heart attacks via real-time analytics on a patient's smartwatch stream [43], for instance).

2.2.3 Privacy and Compliance

In the modern era, as the full ramifications of AI continue to be explored in press and other commentary, lawmakers have started to take note. AI may have major consequences for citizens' rights pertaining to the collection, use, and dissemination of their personal data. We cover many of the major issues in Chap. 5, when we discuss AI ethics and relevant regulations such as the GDPR. Herein, we note that privacy and compliance constitute major challenges to the adoption of AI, especially when a company has to take into account multiple countries and sets of policies.

Interestingly, an argument can also be made in the opposite direction: AI can *help* large companies comply with complex requirements. As one example of how

AI can be used to help with compliance and capital requirements for banks and financial intermediaries, Fraisse and Laporte [7] provide an empirical exercise where they measure the extent to which banks are incentivized to invest in AI to better understand their own capital requirements. Per the Basel II accords (endorsed in 2004), banks can use internal models to estimate the minimum amount of capital they have to hold under law. The model is subject to checks and balances, since it must have prior authorization. Because it is an involved process, the approach is referred to as the "advanced approach" for estimating capital requirements. Without such authorization, banks have to rely on external ratings using the so-called "standardized approach."

Although the full technical details behind the paper's methodology are beyond the scope of the present discussion, the authors found that using machine learning for estimating the credit risk of corporate businesses (which in turn is closely tied to banks' capital requirements) can be more robust than the traditional logistic regression method used by banks to internally model and estimate credit risk for such businesses. The authors also claim that, within reasonable bounds, neither the sample size nor data processing steps such as discretization decisions on continuous data affect stability of the results. However, they do find that there is a loss in accuracy when moving from training to test data, which is not unsurprising. Supervised models can be prone to overfitting issues on training data. Unsurprisingly as well, neural networks were found to provide the strongest benefits.

Ultimately, these results make a compelling case that machine learning models such as neural networks could serve as higher quality internal models that banks can, and should, use to more accurately predict corporate defaults and anticipate capital requirements as a result. There is, of course, a drawback to the use of neural networks, not including the aforementioned overfitting; namely, the latest deep learning models tend to lack explainability, which bank management would likely expect before making major decisions. A lack of explainability may be acceptable for less critical decisions, however. In the future, advances in explainable AI may allow these models to become more mainstream for banks and regulatory agencies alike.

2.2.4 AI Quality Metrics

Businesses tend to measure, and "run on," metrics that have clear connections to financial levers such as revenue, earnings, and cash flow. The financial metrics themselves often influence the business's standing in the stock markets (assuming it is public) or its valuation and are key in executive decision making.

The connection between AI quality metrics and financial metrics is usually not clear and can encounter the famous "chicken-and-egg" paradox: good performance of an AI, evidenced through its quality metrics, can be argued to lead to greater efficiencies and new sources of revenue, but even an approximate estimate would require the AI to be implemented in the first place. However, the decision to implement the AI, especially if cost-prohibitive, requires *some* estimate of how it might impact financial performance. One way to address the paradox is by building and validating forecasting models linking AI quality metrics and financial metrics, but such exercises are still in their infancy in the corporate sphere.

Another problem that is less theoretical, and more social, is that engineers and business executives have good knowledge of their respective metrics but can find it difficult to communicate in a compelling manner how one set of metrics maps on to the other. It is not unusual, then, to find that market pressures (including the "fear of missing out" phenomenon when a rival starts implementing AI, which also then gets covered by the press, to add insult to injury) can often be more effective than internal pressures. Inevitably, winners and losers emerge. Companies that are able to use AI on their own datasets and use cases effectively acquire a first-mover advantage, while others are doomed to playing catch-up.

An even more severe problem is that the choice of metrics for measuring the quality of an AI can sometimes be non-obvious or even under-determined. When measuring supervised machine learning performance, accuracy is the first metric adopted by many, but simple accuracy can be misleadingly high (even when the system is worse than random) due to *data skew*. By simple accuracy, we mean the fraction of correctly classified answers (in the test set) relative to all classified answers. For instance, suppose there were 100 test examples, and 70 of them were classified correctly by the supervised system. The accuracy would then be 70%.

To understand why this metric can be problematic, take the hypothetical case of a consumer-facing company that is using AI to classify negative reviews into two categories: either the (negative) review is related to something intrinsic in the product or service in some way (e.g., it did not work as expected or broke too soon), or the negative review is related to an "auxiliary" or extrinsic experience, such as unexpected shipping delay, difficulty in navigating the website for help, or bad customer service. If the goal is to appease such customers and get them to modify or remove the review, the first kind of negatively reviewing customer may be more easily appeased (perhaps by sending them a new product for free or by issuing them a credit or refund). In contrast, the real value in the second kind of negative review may be to illuminate logistical, or other, problems the company may want to look into *before* the effects start snowballing and lead to a flurry of negative reviews that quickly spread by word of mouth on social media. Such problems, although narrowly defined, are excellent fits for recent advances in Natural Language Processing (NLP) that use advanced neural network models, called transformers, to deliver human-like performance if properly trained.

Suppose, however, that 90% of the negative reviews (say, getting two stars or below) on the platform actually fall into the first category. Simply predicting that *all* negative reviews fall into the first category would then get an accuracy of 90%, but this is hardly illuminating. In such a situation, we caution that accuracy should only be used sparingly, if at all, and a comparison should always be made to a "trivially best" solution, which is very often just a system that predicts randomly or that predicts the "majority" label for all examples.

What metrics can address the skewed situation above? If a single numerical number is desired, as it often is, a favored example in the AI community is the

F1-Score, also called the F1-Measure or F-Measure (for short). We do not provide the technical details or formula behind this score, but its key intuition is that it attempts to trade off two metrics, *precision* and *recall*. Both of these metrics (and by extension, the F1-Score) are always computed with reference to a specific label, called the *positive* label. In the case of the prediction task above, we may designate the first category to be the positive label. Note that "positive" here has nothing to do with the intrinsic definition of the label itself. Instead, it should be thought of as a reference or as the target label of interest for computing the F1-Score.

Similar to simple accuracy, both precision and recall compute fractions. Intuitively, precision measures how many of the positive labels output by the algorithm are actually correct, whereas recall measures how many of the actual positive labels were predicted correctly by the algorithm. An example will prove illustrative, again using the two-class negative review case. Suppose the first class (negative reviews due to defective product or service, hopefully inadvertently) is the reference or the "positive" class. Moreover, in a 100-sample test set, 90% are determined to fall in this positive class by a team of annotators reading the reviews manually and labeling[2] each sample with one of the two labels. Now suppose that, of these 90 *actually correct* samples, the AI system tags 50 as correct. The AI system also incorrectly tags 7 of the other 10 "negative" samples as positive. The recall of such a system would be 50/90 or 55.55%. The precision would be 50/57 or 87.72%. Notice how the numerator is the same in both calculations, but the denominator is different.

The F1-Score is the *geometric average* of precision and recall. We do not state the formula here, but the geometric average is more skewed toward the lower of the two numbers. In the extreme case, if either precision or recall were to be 0%, the F1-Score would necessarily be 0% even if the other metric were to be 100%!

Even with this brief background, it is not difficult to see that these metrics, well understood and established in the AI community, do not have an *obvious* connection to a financial metric like revenue. For instance, even in the narrow and practical negative review case above, how much impact would a 10% improvement in F1-Score have on next quarter's revenues if the roll-out of the improved AI system were instantaneous? What is the minimum F1-Score (e.g., 70%? 90%?) required to have any impact on revenue and is revenue the right business metric to be using as the "dependent" variable? Instead, an intermediate metric like customer retention rate might be a better fit, and since its connection to revenue and earnings is (usually) better understood, the goal of tying in the AI metric to a primary financial metric may be accomplished by going down this more indirect route.

Unfortunately, tying improvements in metrics like F1-Score even to customer retention rates is non-trivial. Directionally, it is theoretically obvious that improve-

---

[2] We assume, for the sake of simplicity, that no other type of negative review categorization is possible. This is a simplifying assumption, but in practice, other "types" of negative reviews, if encountered, could simply be excluded from the test set. Also, note that the test set is only being used to *evaluate* one or more predictive systems. In practice, the team of annotators cannot be used at scale to tag many thousands of reviews, which is why the AI is being developed and tested in the first place.

ments in one direction can only lead to improvements in the other, but whether the effect is dampened or amplified is not easy to determine without an actual experiment. And even if the determination is made with sufficiently high fidelity, it then becomes necessary to determine if the *costs* for achieving that 10% F1-Score improvement can be estimated with equally high fidelity. Much depends here on whether the improvement can be achieved with more manual labor, such as availability of more annotated data, or if a fundamental advance or implementation (such as a new neural network approach) is required, which is inherently riskier. To make matters even worse from a cost perspective, no benefits might be realized without first setting up a complete system or pipeline, which involves engineering effort at all levels.[3]

These problems are difficult, and may seem hopeless, but are not insurmountable. Certainly, companies that spend time on answering such questions are primed for first-mover advantage. We caution against seeking "easy" answers that seem overly certain but are based on faulty assumptions or practices. In the next section, we comment on guidelines for measuring the ROI of AI projects. Similar to how the F1-Score aims to trade off precision and recall in a single number, the ROI is one of the several established ways to trade off the costs and benefits of a project. Its computation for AI projects can be challenging, but with a disciplined corporate approach, is feasible.

2.3   Guidelines and Practices for Measuring Return on Investment (ROI) of AI Projects

When allocating resources to a business venture, the *Return on Investment* (ROI) metric is of obvious importance. Simply defined, ROI is a ratio that quantifies an investment's gain (or loss) relative to cost. Put differently, the benefits of an investment should outweigh the costs by a sufficient margin (depending on the organization) in order for the investment to proceed.

When the venture is primarily a physical investment or otherwise has cost estimates and cash flows that can be predicted with some certainty, standard tools in finance can be used to compute such metrics. When investing in AI, however, calculating ROI can become complex and riddled with uncertainty. For risk-averse organizations, the uncertainty (or in more extreme cases, infeasibility) of computing ROI can doom the project before it can be worked out in detail. As the economy becomes more knowledge-intensive and driven by the types of innovation that require some investment in AI capabilities, there is a need to better understand how to measure ROI of AI projects. A related issue is "valuing" an AI project by placing

---

[3] Indeed, a complete architecture may involve multiple pipelines. As one example, we had mentioned deduplication in the earlier section. In a complete system, we could think of deduplication as a "box" in the system architecture, but this box is likely a pipeline in itself [2, 17].

a dollar amount on it by considering costs and benefits in a net present value or NPV-like framework.

Whichever metric is considered (ROI or NPV), the ultimate issue is one of quantifying costs and benefits in a manner that allows executives or investors to make a decision on whether to proceed with funding the project (or equivalently, investing funds in a company or startup proposing the project). Although research on this continues, in this section, we aim to provide some guidance on the matter through a recently proposed framework by an established professional services firm.

2.3.1 Traditional Valuation Approaches and Their Pitfalls for Valuing AI Projects

As a first step, let us consider the pitfalls when using traditional valuation approaches for AI projects, such as the *market, income*, and *asset* approaches. The market approach involves assessing comparable public companies (colloquially referred to as "comparables"), or "precedent transactions," such as mergers and acquisitions (M&A), to value a private company, or a division of a public company. The market approach has some obvious benefits: it is simple and objective and in practice cannot be "gamed" as easily. Commercial data providers provide platforms on which transactions and public valuation data can also be accessed without much technical effort. The issues arise with the fact that markets may be mis-pricing certain companies due to momentum effects, rumors of impending M&A, or others, and these can skew results when taking averages. Generally, because M&A involve premia, the precedent transactions approach can yield a different (often, higher) valuation compared to the comparables approach. However, such issues are not unique to valuing AI projects; they may arise for any kind of valuation, including other sectors that are not well understood or are new.

The more fundamental issue is that there are no pure AI comparables to begin with in many cases. Precedent transactions are more common in that many Big Tech companies have made a number of "AI acquisitions" in recent years, but the details and transaction amounts may not always be visible. For example, was the company acquired for its personnel (often called an "acqui-hire") or for its technology? More generally, beyond Big Tech, AI is considered synergistic, rather than fundamental, to the products and services offered by companies in sectors such as healthcare and finance. In different industries, different synergies are expected. In the future, as AI matures, using the market approach may become more feasible.

The income approach directly focuses on the future cash flows that will be generated from the project. In typical firms, financial metrics most closely associated with cash flows, such as revenues, net cash flows, and EBITDA (Earnings Before Interest, Tax, Depreciation, and Amortization), can be appropriately discounted to value the project. The initial outlay (and in the same vein, recurring costs) gets treated as a "negative" cash at time t_0 in this framework. When valuing AI projects, it is not only difficult to estimate cash flows but may also be difficult to estimate costs. Given the persistent labor shortage in high-tech, the churn of old technologies, and the costs required to train AI models (whether using in-house servers or the cloud), such as deep neural networks with hundreds of millions, or even billions, of parameters, any estimates are likely to be volatile. Also, as noted earlier, because the benefits of AI are likely to be synergistic, rather than primary, to most companies' product and service offerings, costs and income flows will have to be evaluated in an opportunity cost framework, possibly by invoking counterfactuals and considering not just inaction but non-AI baselines.

The asset approach can be applied to pre-revenue AI projects and can potentially work even when cash flows are negative or minimal, and returns on investment are not expected in the near future. Usually, even statements of financial forecasts cannot be reasonably generated. The approach works by considering the "fair market value" of the relevant assets, net of liabilities. Such assets are generally intangible but can include (for example) the personnel, which is the focus of acqui-hires, sales pipelines, nature and strength of the intellectual property (IP), including patents, and trained AI models that may be difficult to replicate and engineer even if there is no ostensible IP protection.

The asset approach may seem the most promising of the three for generating AI project valuations and ROI, especially for startups. However, determining the fair market value involves a similar set of problems that were earlier noted for the market approach. There is considerable room for subjectivity, which also eliminates one of the strengths we noted for the market approach (that it tends to be difficult to be "gamed" given that there are fewer assumptions and forecasts than some of the other approaches, and it is easier to see where the skew in, or range of, valuation estimates may be stemming from).

2.3.2 Soft Versus Hard Returns and Investments

One way to address the problem of generating an ROI for AI projects is to consider a slightly different definition of ROI, called "soft" ROI. The informal definition of ROI we stated earlier (*direct* benefits of investment outweighing costs by a sufficient margin) can be thought of as "hard" ROI, according to a 2021 article by Anand Rao, who was then global AI lead in the emerging technology group in PricewaterhouseCoopers (PwC), a multinational professional services network of firms. According to him, soft ROI should look, not just at the explicit costs and benefits (as captured by cash flow metrics, for instance) but at a "broader" set of benefits as well. These benefits include, but are not limited to, employee satisfaction and retention, brand enhancement, higher valuation of the company, and advanced skills-acquisition of the employees, which may itself be correlated with employee satisfaction and retention.

The benefits listed above are typically not considered drivers of hard ROI, although there is an argument to be made that they do, in fact, affect it in a non-trivial way. For example, if employees become expensive to retain because they are dissatisfied with morale, hard ROI could plummet. However, the more primary drivers of ROI are easier to measure using more direct metrics, such as productivity increase, revenue increase, and cost savings.

One way to structure the ROI of AI is to first recognize that it can affect the drivers of both hard ROI and soft ROI. AI can increase productivity through process automation, better decision making (including compliance and error checking), and higher efficiency. Improving fundamental offerings such as recommendation services can help some organizations convert trial customers to paid subscribers more easily, as well as to significantly improve retention of existing customers. AI can also yield cost savings, sometimes at the cost of eliminating certain jobs, but more often by reducing errors caused by manual repetition and other consequences of human fatigue. A common case where AI can assist humans rather than replace them is by automatically flagging errors in data entry that may otherwise not get caught until they have had an unintended consequence (or manifested in a downstream product or service), thereby averting potentially catastrophic consequences, and many lost hours in productivity and morale.

In terms of affecting soft ROI, AI can help data science and engineering teams be more agile and even become innovators. Skills retention is becoming increasingly important in a knowledge-driven economy. By incentivizing its workers to learn and apply AI skills, even if the actual application does not manifest in a consumer facing product or service, the company is also prepared to weather future competition or black swan events that require a paradigmatic shift to digital (as evidenced by COVID-19).

According to Rao, a company should also review its AI expenditures (or investments) through a hard–soft lens. He includes as soft investments the following:

1. **Data investments:** We noted earlier that data acquisition and quality are important challenges to address in any ambitious implementation of AI. Availability and accessibility of labeled data, including the ease with which labels can be acquired and shared, are equally critical drivers. It is important for companies to be realistic, and not underestimate, the data challenges and investments in implementing AI (and machine learning models, in particular).
2. **Compute and storage investments:** One of the costs that is easier to foresee is the cost of storage and computation ("compute"). However, storage and compute can be irregular, and the on-demand provisioning that cloud computing allows for can also rake up huge bills for the careless. It is important to budget appropriately for the computational needs of a project and to allow for a safe margin. Computational requirements become even more important when the company starts to implement models on Big Data, especially involving deep learning (and other such complex) architectures.
3. **Investments in domain expertise:** Domain (also known as *subject matter*) experts are necessary in specific applications to ensure that the AI is being built with real users and use cases in mind. However, domain experts' primary job is not in assisting AI modelers and computer scientists. The "time investments" required for domain experts should therefore be carefully estimated and budgeted for. It should be recognized by the team leaders which phases of an AI project

such experts will be critically needed for. For truly ambitious projects, such experts tend to be required (although not equally critically) in all phases of an AI project, from scoping and building the initial models to the deployment and monitoring of those models in actual practice. If a company cannot guarantee the availability of these experts during these phases, the timeline and planning for the AI project should be appropriately adjusted.

4. **Investing in data science and other relevant skills training:** An extremely important enabler of rapid adoption of AI and data science in many sectors is a dynamic and vibrant ecosystem of academics, open-source datasets and software packages, openly published and disseminated research, and a community of practitioners (hailing from academia, industry, government, and even middle- and high-schools). Companies must invest in providing an equally vibrant environment for training, coaching, and mentoring their employees so that they are equipped with the latest tools, research, and best practices. Needless to say, this may require a cultural shift in the company. It is not a stretch to imagine that all companies will indeed have to become "technology" companies to facilitate such a culture throughout the company. Even more ambitious and forward thinking companies will see to it that *all* of their employees be given a foundational education in AI, data science, and statistics so that their opportunities, challenges, and benefits can be recognized and taken advantage of, from the bottom up.

Given that investments and returns can both be framed using the hard–soft lens, a 2 × 2 matrix can qualitatively capture the drivers that a company should consider when thinking about valuing an AI project. We reproduce this matrix in Fig. 2.1, with the caveat that, although the items in the four "buckets" are typical items, they may be different for different companies. Larger organizations will likely want to engage with a broad group of internal stakeholders, including directors and middle management, and possibly hire consultants, if investing in the AI project is a major and strategic undertaking. For smaller projects, involving only a few teams, the exercise could probably be done by the team managers themselves, with a director overseeing it to maintain objectivity.

Finally, to obtain a numerical ROI, the different drivers need to be factored into the ROI formula, which is simply defined as *Return/Investments*. *Return* can further be formulated as the benefits from the AI project *net* the uncertainty of the benefits. When considering such benefits, engineers should be consulted to understand both the predictive power, and errors, of any model that is being implemented. A more business-facing stakeholder or product manager may then have to be consulted to understand the values of those predictions (which factor into the benefits) and the costs and impacts of the errors (which factor into the uncertainty of the benefits). This may be a non-trivial problem, compounded if the models are based, not on supervised learning that directly yields predictions, but unsupervised or reinforcement learning. Either way, it is an unavoidable step. Recall, for instance, in the earlier section on AI quality metrics, the challenge (at times) of converting to business metrics. We note furthermore that diversity of thought and background is always beneficial when making such translations.

Fig. 2.1 A 2 × 2 matrix based on a proposal discussed informally in [35], expressing the "hard" and "soft" drivers of investments and returns for AI projects

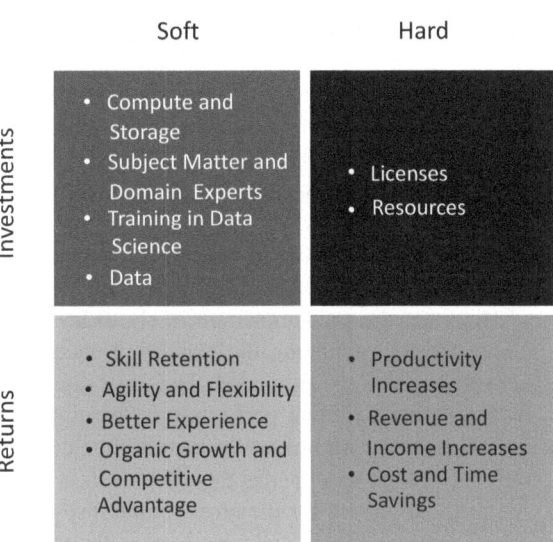

Using a similar methodology, *investments* can be formulated as the product of the resources to build such a model and the cost of those resources. Finally, organizations could choose to compute hard and soft ROI separately, with the understanding that soft ROI may be more uncertain, but possibly have longer term and systemic benefits that should be borne in mind. A weighting scheme could be adopted to consider the tradeoff between hard and soft ROI. A company's culture of innovation and risk-tolerance, as well as the economic cycle itself, will play a role in the choice of these weights. During times when the company is flush with capital and more prone to engage in projects with low downside risk for its existing revenue-generating offerings, but high upside potential, it may actually end up preferring a project with high soft ROI (with high error bounds) compared to one with lower hard ROI (with smaller error bounds). The economic cycle matters because the ROI threshold for undertaking a new project may be high in times of recession, and a greater premium may be placed by the company on hard ROI, with clear projected benefits and less uncertainty than the more uncertain (and hence, riskier) soft ROI.

We also note that, although converting the qualitative matrix to quantitative estimates and forecasts may seem intimidating initially, it is no different (at least, in principle) than making disciplined forecasts for other projects being valued. Rao does offer certain caveats that are specific to AI and should be borne in mind when making forecasts:

1. **Discounting uncertainty of benefits:** Although many organizations rely on simple ROI calculations for their AI projects, they do not adequately quantify or estimate the uncertainty that is associated with realizing the benefits. An example that Rao uses is an AI system that takes as input, a customer complaint as free-form text, and outputs as a prediction whether the complaint is of high, medium,

or low severity. This problem, which is also known as *urgency detection*, is also relevant in domains such as crisis informatics, where urgent situations (especially during natural disasters) on social media need to be quickly flagged for action or review [28–30]. Within the ROI framework just presented, the return of such a project can be estimated by first estimating the value of each prediction made by the AI, and the number of such predictions made within a period of time (say, annually). Most likely, the value is directly correlated to the number of minutes saved by a customer service representative in moving from a (previously) manual to such an AI-assisted, automated solution.

However, the predictions are likely to be noisy; hence, there is uncertainty associated with realizing the benefits. Hence, both the error rate and the cost per error need to be estimated. The error rate could be computed, for instance, by evaluating a human performance "baseline" with the AI's prediction error rate. Note that it would not be wise to just consider the AI's error rate without reference to the baseline, as this will result in an overestimate of the AI's costs. A statistical margin for the error should also be estimated, since the real-world deployment will likely yield more errors than the "clean" environment witnessed during training (even if every attempt was made to replicate the real world).

Computing the cost per error is where the real challenge emerges. There is obviously a different cost for mis-labeling complaints of different severity. Erroneously tagging a high-severity complaint as a low-severity complaint seems to intuitively incur the highest cost, but there is also a hefty opportunity cost to the company, including wasted resources, when a low-severity message is tagged as a high-severity message. While data can help to estimate true costs over time, an important challenge that many companies face is that they rarely have the infrastructure or mindset to capture data in a way that would allow them to make these estimates. Estimates may therefore be nothing short of guesses, with wide error margins. In turn, this may lead to disappointing results and abandonment of the AI project when times become lean (such as in a recession).

2. **Computing ROI only for a specific point in time:** Another caveat is to avoid computing the ROI of an AI project only at a specific point in time. Usually, this time is a few months, or a short period, after the AI system has been deployed into production. It is well known, unfortunately, that adaptive models such as those based on supervised machine learning may potentially deteriorate over time due to a variety of technical issues. One such issue is *concept drift* wherein the data and task are slowly changing over time [8, 44] and eventually become different enough from the original training data, and that the error rates start worsening. Therefore, AI performance should be measured or audited on a continuous basis, and the costs of such auditing should be taken into account when planning the project in the first place. Adequate "maintenance cost" budgeting should also be implemented to "repair" the model when its performance dips below a certain level, so that the value of the system does not get eroded by the benefits already achieved. In the long term, such maintenance costs can add up, but the model also becomes difficult for competitors to copy. Done correctly, the model may end up being a source of major competitive advantage if continuously improved and maintained, such as Google's search engine and Netflix's recommender system.

3. **Treating each AI project individually:** A final caveat noted by Rao is that companies do not consider the synergies between AI projects and tend to treat each AI project individually. While this approach may work for physical investments, such as setting up a factory or launching a new product, synergies are extremely important in business. A more M&A approach needs to be taken when considering the launch of a new AI project. AI projects should be viewed as components of a holistic portfolio, and the goal should always be to maximize the ROI of the portfolio as a whole. In turn, this requires more systems level, as opposed to reductionist, thinking. It may also require collaborations between the different teams that take ownership of the different AI projects.

We conclude this section with a note on how soft and hard ROI are expected to influence decision making in industries of the future. Initially, we believe that there will continue to be a focus on hard ROI, especially in "physical" industries like manufacturing, 3D printing, or even biotechnology. Inevitably, as prices normalize, and physical capital diffuses throughout the industry, the focus will shift to innovation and differentiation. More attention should be paid to soft ROI at this inflection point, and the success of many companies may well depend on whether they are able to identify, and capitalize on, this inflection point in their lifecycle before their competitors. In contrast, in certain industries, including Big Tech, soft ROI is already a predominant metric, although it may not be referred to with that moniker by the company's top brass. The importance of intangibles, the dependence on human capital, and IP, all speak to the prevalence of soft ROI in these companies in an indirect way, however. For those companies, building more accurate models of soft ROI and discovering company-specific drivers that can be measured, quantified, and tuned to obtain the most value out of expensive data and AI investments will become the priority.

2.4   Digital Technology and the Productivity Puzzle

Although ROI and cost-benefit analysis of AI projects inevitably occur at the level of individual firms, the analysis also begs a larger question: where, or when, will the productivity payoff be achieved from emerging technologies such as AI? In the New York Times [32], an article by reporter Steve Lohr explores this so-called *productivity puzzle*, which is stated as the "mystery" that, despite such enormous strides in such digital technologies, annual productivity growth in almost every industrialized economy has been quite low, inching along at less than (or just above) 1% per annum, compared to the 3% productivity growth (or above) that was experienced in the period from the mid-1990s to the mid-2000s. And indeed, even that was slower than the 3.8% productivity growth that was seen during the post-war period between 1947 until the early 1970s, when the economy experienced stagflation due to the oil shock (among other reasons). As with many economic puzzles, there are two leading arguments. The first is by noted economist Robert Gordon, who is skeptical of productivity gains from such technologies. Gordon continues to maintain that much of the productivity growth we are going to see in the foreseeable future has already been achieved in the post-war period and that the new technology, while "impressive," is not "transformational" from a productivity perspective [9, 10]. On the other side of the argument is Erik Brynjolfsson, who directs Stanford University's Digital Economy Lab and who believes that the impact of AI (in particular) *is* transformational, and that we will soon see productivity gains to the tune of 1.8% on average in the decade to come.[4] Whose claims will prevail is still a matter of debate. We believe that the next ten years will see some impact from AI, but that the true industries of the future are along a further and more uncertain time horizon (i.e., anywhere in the range of the next 5-30 years, rather than the next decade).

2.5   Conclusion

We started with the question: given that AI has the potential to transform industry, what challenges, issues, and costs have stood in the way of its implementation? Although a full consensus does not exist, there is some agreement on the technical challenges that stand in the way of a robust and company-wide implementation. This chapter discussed some of these challenges, which include issues related to data acquisition, data quality, privacy and compliance, and perhaps most importantly, devising and aligning good AI quality metrics with financial and business metrics.

In the second half of this chapter, we considered the important issue of balancing costs and benefits. Every new technology and its implementation involves both a cultural shift and an implementation cost. Companies have indeed been taking on some of these costs, mostly out of a fear of being left behind by others in their industry. There is less focus, especially beyond Big Tech, of using rigorous ROI measures for evaluating the value proposition of an AI project. Even when such measures are adopted, our suspicion is that there is too much focus on "hard" ROI, which may not provide the most accurate picture of the cost-benefit tradeoff that applies to an AI project. Alongside hard ROI, therefore, there also needs to be a focus on "soft" ROI.

References

1. Albert, P.S.: Longitudinal data analysis (repeated measures) in clinical trials. Statistics in medicine **18**(13), 1707–1732 (1999)
2. Balaji, J., Javed, F., Kejriwal, M., Min, C., Sander, S., Ozturk, O.: An ensemble blocking scheme for entity resolution of large and sparse datasets. CoRR **abs/1609.06265** (2016). URL http://arxiv.org/abs/1609.06265

---

[4] The two have even made a 400$ bet that will be settled in 2029 when the data is compiled to verify this claim.

3. Bandt, C.: Autocorrelation type functions for big and dirty data series. arXiv preprint arXiv:1411.3904 (2014)
4. Cowie, M.R., Blomster, J.I., Curtis, L.H., Duclaux, S., Ford, I., Fritz, F., Goldman, S., Janmohamed, S., Kreuzer, J., Leenay, M., et al.: Electronic health records to facilitate clinical research. Clinical Research in Cardiology 106(1), 1–9 (2017)
5. Deng, J., Dong, W., Socher, R., Li, L.J., Li, K., Fei-Fei, L.: ImageNet: A large-scale hierarchical image database. In: 2009 IEEE conference on computer vision and pattern recognition, pp. 248–255. IEEE (2009)
6. Desmond, M., Duesterwald, E., Brimijoin, K., Brachman, M., Pan, Q.: Semi-automated data labeling. In: NeurIPS 2020 Competition and Demonstration Track, pp. 156–169. PMLR (2021)
7. Fraisse, H., Laporte, M.: Return on investment on artificial intelligence: the case of bank capital requirement. Journal of Banking & Finance p. 106401 (2022)
8. Gama, J., Žliobaitė, I., Bifet, A., Pechenizkiy, M., Bouchachia, A.: A survey on concept drift adaptation. ACM computing surveys (CSUR) 46(4), 1–37 (2014)
9. Gordon, R.J.: Is us economic growth over? faltering innovation confronts the six headwinds. Tech. rep., National Bureau of Economic Research (2012)
10. Gordon, R.J.: Why has economic growth slowed when innovation appears to be accelerating? Tech. rep., National Bureau of Economic Research (2018)
11. Hoerbst, A., Ammenwerth, E.: Electronic health records. Methods of information in medicine 49(04), 320–336 (2010)
12. Hopgood, A.A.: Artificial intelligence: hype or reality? Computer 36(5), 24–28 (2003)
13. Jobin, A., Ienca, M., Vayena, E.: The global landscape of ai ethics guidelines. Nature Machine Intelligence 1(9), 389–399 (2019)
14. Kang, J.K., Stice-Lawrence, L., Wong, Y.T.F.: The firm next door: Using satellite images to study local information advantage. Journal of Accounting Research 59(2), 713–750 (2021)
15. Kejriwal, M.: Populating entity name systems for big data integration. In: P. Mika, T. Tudorache, A. Bernstein, C. Welty, C.A. Knoblock, D. Vrandecic, P. Groth, N.F. Noy, K. Janowicz, C.A. Goble (eds.) The Semantic Web - ISWC 2014 - 13th International Semantic Web Conference, Riva del Garda, Italy, October 19–23, 2014. Proceedings, Part II, *Lecture Notes in Computer Science*, vol. 8797, pp. 521–528. Springer (2014). DOI URL https://doi.org/10.1007/978-3-319-11915-1_34
16. Kejriwal, M.: Entity resolution in a big data framework. In: B. Bonet, S. Koenig (eds.) Proceedings of the Twenty-Ninth AAAI Conference on Artificial Intelligence, January 25–30, 2015, Austin, Texas, USA, pp. 4243–4244. AAAI Press (2015). URL http://www.aaai.org/ocs/index.php/AAAI/AAAI15/paper/view/9294
17. Kejriwal, M., Liu, Q., Jacob, F., Javed, F.: A pipeline for extracting and deduplicating domain-specific knowledge bases. In: 2015 IEEE International Conference on Big Data (IEEE BigData 2015), Santa Clara, CA, USA, October 29–November 1, 2015, pp. 1144–1153. IEEE Computer Society (2015). DOI URL https://doi.org/10.1109/BigData.2015.7363868
18. Kejriwal, M., Miranker, D.P.: An unsupervised algorithm for learning blocking schemes. In: H. Xiong, G. Karypis, B. Thuraisingham, D.J. Cook, X. Wu (eds.) 2013 IEEE 13th International Conference on Data Mining, Dallas, TX, USA, December 7–10, 2013, pp. 340–349. IEEE Computer Society (2013). DOI URL https://doi.org/10.1109/ICDM.2013.60
19. Kejriwal, M., Miranker, D.P.: On linking heterogeneous dataset collections. In: M. Horridge, M. Rospocher, J. van Ossenbruggen (eds.) Proceedings of the ISWC 2014 Posters & Demonstrations Track a track within the 13th International Semantic Web Conference, ISWC 2014, Riva del Garda, Italy, October 21, 2014, *CEUR Workshop Proceedings*, vol. 1272, pp. 217–220. CEUR-WS.org (2014). URL http://ceur-ws.org/Vol-1272/paper_17.pdf
20. Kejriwal, M., Miranker, D.P.: A two-step blocking scheme learner for scalable link discovery. In: P. Shvaiko, J. Euzenat, M. Mao, E. Jiménez-Ruiz, J. Li, A. Ngonga (eds.) Proceedings of the 9th International Workshop on Ontology Matching collocated with the 13th International Semantic Web Conference (ISWC 2014), Riva del Garda, Trentino, Italy, October 20, 2014, *CEUR Workshop Proceedings*, vol. 1317, pp. 49–60. CEUR-WS.org (2014). URL http://ceur-ws.org/Vol-1317/om2014_Tpaper5.pdf

21. Kejriwal, M., Miranker, D.P.: Decision-making bias in instance matching model selection. In: M. Arenas, Ó. Corcho, E. Simperl, M. Strohmaier, M. d'Aquin, K. Srinivas, P. Groth, M. Dumontier, J. Heflin, K. Thirunarayan, S. Staab (eds.) The Semantic Web - ISWC 2015 - 14th International Semantic Web Conference, Bethlehem, PA, USA, October 11–15, 2015, Proceedings, Part I, *Lecture Notes in Computer Science*, vol. 9366, pp. 392–407. Springer (2015). DOI URL https://doi.org/10.1007/978-3-319-25007-6_23

22. Kejriwal, M., Miranker, D.P.: A DNF blocking scheme learner for heterogeneous datasets. CoRR **abs/1501.01694** (2015). URL http://arxiv.org/abs/1501.01694

23. Kejriwal, M., Miranker, D.P.: Minimally supervised instance matching: An alternate approach. In: F. Gandon, C. Guéret, S. Villata, J.G. Breslin, C. Faron-Zucker, A. Zimmermann (eds.) The Semantic Web: ESWC 2015 Satellite Events - ESWC 2015 Satellite Events Portorož, Slovenia, May 31–June 4, 2015, Revised Selected Papers, *Lecture Notes in Computer Science*, vol. 9341, pp. 72–76. Springer (2015). DOI URL https://doi.org/10.1007/978-3-319-25639-9_14

24. Kejriwal, M., Miranker, D.P.: On the complexity of sorted neighborhood. CoRR **abs/1501.01696** (2015). URL http://arxiv.org/abs/1501.01696

25. Kejriwal, M., Miranker, D.P.: Semi-supervised instance matching using boosted classifiers. In: F. Gandon, M. Sabou, H. Sack, C. d'Amato, P. Cudré-Mauroux, A. Zimmermann (eds.) The Semantic Web. Latest Advances and New Domains - 12th European Semantic Web Conference, ESWC 2015, Portoroz, Slovenia, May 31–June 4, 2015. Proceedings, *Lecture Notes in Computer Science*, vol. 9088, pp. 388–402. Springer (2015). DOI URL https://doi.org/10.1007/978-3-319-18818-8_24

26. Kejriwal, M., Miranker, D.P.: Sorted neighborhood for schema-free RDF data. In: F. Gandon, C. Guéret, S. Villata, J.G. Breslin, C. Faron-Zucker, A. Zimmermann (eds.) The Semantic Web: ESWC 2015 Satellite Events - ESWC 2015 Satellite Events Portorož, Slovenia, May 31–June 4, 2015, Revised Selected Papers, *Lecture Notes in Computer Science*, vol. 9341, pp. 217–229. Springer (2015). DOI URL https://doi.org/10.1007/978-3-319-25639-9_38

27. Kejriwal, M., Miranker, D.P.: Sorted neighborhood for schema-free RDF data. In: J. Völker, H. Paulheim, J. Lehmann, V. Svátek (eds.) Proceedings of the 4th Workshop on Knowledge Discovery and Data Mining Meets Linked Open Data co-located with 12th Extended Semantic Web Conference (ESWC 2015), Portoroz, Slovenia, May 31, 2015, *CEUR Workshop Proceedings*, vol. 1365. CEUR-WS.org (2015). URL http://ceur-ws.org/Vol-1365/preface.pdf

28. Kejriwal, M., Zhou, P.: Low-supervision urgency detection and transfer in short crisis messages. In: F. Spezzano, W. Chen, X. Xiao (eds.) ASONAM '19: International Conference on Advances in Social Networks Analysis and Mining, Vancouver, British Columbia, Canada, 27–30 August, 2019, pp. 353–356. ACM (2019). DOI URL https://doi.org/10.1145/3341161.3342936

29. Kejriwal, M., Zhou, P.: Low-supervision urgency detection and transfer in short crisis messages. CoRR **abs/1907.06745** (2019). URL http://arxiv.org/abs/1907.06745

30. Kejriwal, M., Zhou, P.: On detecting urgency in short crisis messages using minimal supervision and transfer learning. Soc. Netw. Anal. Min. **10**(1), 58 (2020). DOI URL https://doi.org/10.1007/s13278-020-00670-7

31. Kusiak, A.: Smart manufacturing. International Journal of Production Research **56**(1-2), 508–517 (2018)

32. Lohr, S.: Why isn't new technology making us more productive? (2022). URL https://www.nytimes.com/2022/05/24/business/technology-productivity-economy.html

33. McKendrick, J.: Ai adoption skyrocketed over the last 18 months (2021). URL https://hbr.org/2021/09/ai-adoption-skyrocketed-over-the-last-18-months

34. Montani, S., Striani, M.: Artificial intelligence in clinical decision support: a focused literature survey. Yearbook of medical informatics **28**(01), 120–127 (2019)

35. Rao, A.: Solving AIs ROI problem. It's not that easy. (2021). URL https://www.pwc.com/us/en/tech-effect/ai-analytics/artificial-intelligence-roi.html

36. Sergi, B.S., Popkova, E.G., Bogoviz, A.V., Litvinova, T.N.: Understanding industry 4.0: AI, the internet of things, and the future of work. Emerald Group Publishing (2019)

37. Sidi, F., Panahy, P.H.S., Affendey, L.S., Jabar, M.A., Ibrahim, H., Mustapha, A.: Data quality: A survey of data quality dimensions. In: 2012 International Conference on Information Retrieval & Knowledge Management, pp. 300–304. IEEE (2012)
38. Slota, S.C., Fleischmann, K.R., Greenberg, S., Verma, N., Cummings, B., Li, L., Shenefiel, C.: Good systems, bad data?: interpretations of AI hype and failures. Proceedings of the Association for Information Science and Technology 57(1), e275 (2020)
39. Tanner, A.: Our bodies, our data: how companies make billions selling our medical records. Beacon Press (2017)
40. Thomas, R.L., Uminsky, D.: Reliance on metrics is a fundamental challenge for AI. Patterns 3(5), 100476 (2022)
41. Thorp, J.: Big data is not the new oil (2012). URL https://hbr.org/2012/11/data-humans-and-the-new-oil
42. Tian, A., Kejriwal, M., Miranker, D.P.: Schema matching over relations, attributes, and data values. In: C.S. Jensen, H. Lu, T.B. Pedersen, C. Thomsen, K. Torp (eds.) Conference on Scientific and Statistical Database Management, SSDBM '14, Aalborg, Denmark, June 30–July 02, 2014, pp. 28:1–28:12. ACM (2014). DOI URL https://doi.org/10.1145/2618243.2618248
43. Tison, G.H., Sanchez, J.M., Ballinger, B., Singh, A., Olgin, J.E., Pletcher, M.J., Vittinghoff, E., Lee, E.S., Fan, S.M., Gladstone, R.A., et al.: Passive detection of atrial fibrillation using a commercially available smartwatch. JAMA cardiology 3(5), 409–416 (2018)
44. Zhang, S., Kejriwal, M.: Concept drift in bias and sensationalism detection: an experimental study. In: F. Spezzano, W. Chen, X. Xiao (eds.) ASONAM '19: International Conference on Advances in Social Networks Analysis and Mining, Vancouver, British Columbia, Canada, 27-30 August, 2019, pp. 601–604. ACM (2019). DOI URL https://doi.org/10.1145/3341161.3343690
45. Zuchowski, O., Posegga, O., Schlagwein, D., Fischbach, K.: Internal crowdsourcing: conceptual framework, structured review, and research agenda. Journal of Information Technology 31(2), 166–184 (2016)

AI in Industry Today

3

3.1   Introduction

As the old saying goes, we cannot know where we are going unless we know where we are coming from. In keeping with this wisdom, the best way to learn about how AI in industry will evolve in the future is by learning from the present, which itself is a convergence of factors that have their origins over multiple decades in the past. These factors are complex, and their full (and long-term) ramifications are still being debated, but some of the impacts are already starting to be witnessed in government and private life alike. Technological factors include the proliferation and lowering of cost (at least when controlled for quality) of key enabling "platforms" such as the social media giants, search engines such as Google, e-commerce behemoths such as Amazon, and on the hardware front, smartphones and tablets. We refer to these innovations as platforms because they are often necessary for other downstream innovations to percolate and reach critical mass. Indeed, Search Engine Optimization (SEO) continues to be important for many businesses, and one can even imagine the work of modern influencers on social media platforms to be an organic, more content-driven version of SEO in the attention economy.

At the same time, while the impact of the so-called Big Tech companies should not be underestimated, it is also easy for the pendulum to swing too far the other way, and for technologists to underestimate the impact of *other* companies on AI adoption and innovation. Our view is that adoption and innovation are actually intertwined in AI, rather than a linear path as much of applied science has historically followed. Later in this chapter, we provide striking evidence of this when we discuss the emergence of large-scale "language models" in the natural language processing community. These models have become remarkably good at generating and interpreting human language, with numerous applications beyond Big Tech. Because of their large scale, industry (rather than academia) is leading the way in training such models, and releasing pre-trained versions of these models for broader use. In some cases, the models are exposed via an application programming

© The Author(s), under exclusive license to Springer Nature Switzerland AG 2023
M. Kejriwal, *Artificial Intelligence for Industries of the Future*, Future of Business and Finance, https://doi.org/10.1007/978-3-031-19039-1_3

interface (API) that requires payment. The key point to note here is that adoption and innovation in this important subfield of AI already seem to be locked in a virtuous cycle.

In this chapter, we aim to take a broad view of AI in industry today, with the overarching goal of extrapolating these trends to adoption and innovation in AI in industries of the future. Some of the industries of the future will be the direct consequence of the evolution of big companies today, whether classified as Big Tech or not. Others may be the startups of today and the Big Techs of tomorrow. The success of specific *companies* is not for us to predict, but we can use the trends we observe to comment (hopefully with more accuracy) on the potential for AI in specific *industries* based on today's observations. Because the role of startups, and small- and medium-sized enterprises (SMEs), is so important for understanding where the future is headed, we give some attention to what we are observing in that segment of industry today, despite their obvious volatility and high probability (based purely on historical grounds) of failure.

3.2   AI in Big Tech

No discussion of AI in the present era can afford to ignore either the influence or contributions of Big Tech. It can sometimes be confusing (with good reason) which companies are considered Big Tech, and which are not. The reason for the confusion can perhaps be traced pithily to the fact that almost every company today is a "technology company." In a sense, then, every "big" company that heavily draws on, or uses, technology (and almost all do) might be designated as Big Tech. Clearly, that is not the case, nor should it be, as the moniker would become meaningless if we were to assign a "literal" meaning to it.

Simply put, there are historical and market-based reasons why a company has been designated as Big Tech in popular discourse; it is a name that has "stuck," just like the "Big Four" in the accounting industry [Deloitte, Ernst & Young (EY), PricewaterhouseCoopers (PwC), and Klynveld Peat Marwick Goerdeler (KPMG)]. Industry watchers agree that the Big Tech firms include Alphabet (Google), Apple, Amazon, Meta (Facebook), and Microsoft. Broader groupings may also include companies such as Netflix and Twitter, but these companies are smaller than the "Big Five," and have less impact on the overall economy or even just AI innovations (more narrowly).

Big Tech has also been described by colorful acronyms, although they are now starting to fall out of favor due to the re-branding of companies such as Facebook (to Meta). One such acronym, coined in 2013 by Jim Cramer, the colorful television host of CNBC's Mad Money, was FANG (Facebook, Amazon, Netflix, and Google). At the time, these companies were quoted by Cramer as being "totally dominant" in the markets in which they operated. In a testament to how quickly markets move and adopt, and how non-Big Tech have now started to compete increasingly with Big Tech in some of these markets (and vice versa), Netflix is now facing extraordinary competition in the ongoing and aptly dubbed "streaming wars." At the time of writing, in July 2022, its stock is down more than 65% compared to just a year earlier.

FANG was expanded to FAANG (by Cramer himself) in 2017, with the extra "A" used to designate Apple. In late 2021 came the recognition that markets have shifted and new winners and losers may be emerging. Microsoft, for example, had met with unexpected success in over the second half of the last decade (in particular) under Satya Nadella's leadership, and re-setting of priorities. Cramer proposed in October of that year to substitute FAANG with MAMAA, a clever play that recognized the change in names since FAANG was proposed: Google was now Alphabet, and Facebook had recently changed its name to Meta to reflect its own priorities in the virtual reality space aka *metaverse*.[1]

3.2.1   Alphabet

As noted above, Alphabet is largely synonymous with Google, despite the rechristening. The now ubiquitous search engine uses a surprising amount of AI to deliver its search results and also advertise effectively. However, while Google may be the keystone company within the Alphabet umbrella (and accounts for much of its revenue and profits), there are other firms within Alphabet that are making intriguing contributions to AI, and beyond.

Indeed, according to some industry watchers, including Crunchbase, Google had acquired 30 or so startups between 2009 and 2020, making it one of the most prolific acquirers of AI startups (to the tune of almost $4B spent on such acquisitions). Examples of companies acquired by Google include PittPatt, Onward, and DeepMind. Of these, DeepMind has become an iconic name in the modern deep learning community, having bested human champions in games such as Go and Shogi, and more recently, achieving groundbreaking performance in practical domains such as bioinformatics. DeepMind's success has largely been due to its pioneering work in the field of *deep reinforcement learning*.

A case study of such a success is the recently released AlphaFold system [66]. AlphaFold and other previous efforts like it are motivated by the fact that experimental determination of protein structures can be time-consuming. With AlphaFold, it was shown for the first time that AI could accurately (at the level of atomic accuracy) and scalably (within minutes) predict the shape of a protein. Previous systems had fallen far short of achieving even 50% median free-modeling accuracy, a metric used to quantify success on this task. In 2018, the first version of AlphaFold entered into a competition (the Community Wide Experiment on the Critical Assessment of Techniques for Protein Structure Prediction or CASP) was just shy of the 60% mark; however, the most recent AlphaFold 2.0 system has cut that margin of error by half. Before this had been accomplished, many had believed

---

[1] Needless to say, Netflix has been unceremoniously dropped from the club in this new acronym.

that the milestone was a half-century away, attesting to the predictive potential of these kinds of deep learning architectures.

In the academic community, Google has also become well known for its recent successes in building, and releasing pre-trained versions of, large-scale *language models* (more technically, language *representation* models). These models, which have enormously impacted natural language processing (NLP) , and are now being applied in other AI fields, are based on a class of deep neural networks called *transformers*. Intuitively, these neural networks are capable of elegantly handling long-range dependencies, which is important when parsing text. However, this is not the only reason why they work so well, since other neural networks before them also had this ability. Indeed, the main reason why transformers are believed to be empirically superior to previous generations of language models is due to their incorporating a novel, and appropriately named, concept called *attention* into their learning procedure.

Although the theory of attention (in the neural network, not cognitive, sense), and the full suite of reasons why transformers work so well, is highly technical and still being worked out in the foundational machine learning community, there is ample demonstration already that, on tasks ranging from question answering to poetry writing, they can be eerily human-like. New and ever larger generations of language models get released every year, and in the United States, Big Tech is at the forefront of these releases. In large part, this is due to their ability to bear the high computational costs.

Pragmatic use of language models in industry, and their influence on NLP (and other fields and applications in AI that are NLP-adjacent), will be the focus of a subsequent case study in this chapter; herein, we note that there is already documented evidence that the famous Bidirectional Encoder Representations from Transformers (BERT) model has already been integrated into the Google search engine. According to a Google blog post by Pandu Nayak, who at the time of writing the blog was Google Fellow and the Vice President of Search:

"... *when it comes to ranking results, BERT will help Search better understand one in 10 searches in the U.S. in English, and we'll bring this to more languages and locales over time.*

Particularly for longer, more conversational queries, or searches where preposi-tions like 'for' and 'to' matter a lot to the meaning, Search will be able to understand the context of the words in your query. You can search in a way that feels natural for you."

If the statement contains accurate information, one cannot understate the prac-tical impact that "better" understanding of one in ten searches would have, given that advertising and search still account for the lion's share of Google's revenue and profits. Furthermore, search is becoming more complex as people type in longer queries, even full sentences and questions, with the expectation that the query will be appropriately "understood" and serviced by the search engine.

3.2.2   Amazon

Beyond its obvious e-commerce and logistics footprint, Amazon may be better known among industry watchers for its cloud computing platform (Amazon Web Services), and its non-AI acquisitions, such as Whole Foods and (more recently) an $8.5B acquisition of MGM studio, but it is also prominent in the AI space. In part, this is necessary given its wide catalog of products, which requires an emphasis on search and recommendation. In the case of Amazon, the search and recommendation problem is likely even more dynamic than it was for traditional retailers, due to it being the marketplace of choice for many third-party sellers.

Two examples of innovations in Amazon that rely heavily on AI, one of which is highly public-facing (and the other, more "backend") are Amazon Alexa and the Amazon Product Graph. Based on a Polish speech synthesizer named Ivona and acquired in 2013 by Amazon, Alexa is a virtual assistant that has now been integrated across the Amazon ecosystem, including the Amazon Echo smart speaker. Alexa includes capabilities such as music playback, voice interactions, streaming podcasts, and providing real-time information, including news, weather, sports, and traffic. Similar to Nest and other home automation systems, Alexa is also capable of functioning as a home automation system.[2] Impressively, Alexa's capabilities can also be extended by users themselves by uploading "skills" (e.g., customized weather programs) into the platform. The AI tools and techniques that seem[3] to be employed by Alexa include, but are likely not limited to, NLP, automatic speech recognition, knowledge graphs, as well as (potentially) rule-based AI to handle common commands and corner cases.

There is no doubt in any observer's mind that Amazon is bullish on Alexa. Even five years ago, more than 10,000 employees were working on Alexa and related products, and in early 2019, Amazon's devices team announced that at least 100 million Alexa-enabled devices had already been sold. Today, Alexa is fully capable of performing a variety of preset functions straight out of the box. These functions include (in addition to the skills and capabilities noted earlier) creating lists and accessing Wikipedia articles. More recently (in June 2022), a controversial feature was announced but is not public-facing yet, whereby Alexa could speak to a user in the voice of a deceased relative, which make some feel "uneasy," to quote NPR [4].

To use or alert an Alexa-enabled device to follow a function-command, a user simply has to say a designated "wake word" (the default being "Alexa"). Alexa is listening for the command following the wake word and then executes the proper

---

[2] Indeed, in early 2019, Apple, Amazon, Google, and Zigbee Alliance announced a partnership to make smart home products more compatible, not dissimilar to how standards are in place (sometimes due to regulation) for other longstanding technologies, such as cars and refrigerators.

[3] Similar to the Google search engine, and other such systems, all the details behind the architecture of such complex products and services are not known and must be extrapolated given information such as publications, patents, public interviews with (and presentations by) Alexa researchers, and journalistic reporting, as well as our current knowledge about the state-of-the-art in fields such as natural language processing.

function. The general workflow is that speech recognition first has to be used to transcribe the user's sound waves into text (a functionality that is more visible in, and not dissimilar to, the text transcriptions that Zoom makes available after meetings to enterprise customers, for instance). Following this step, other AI fields such as NLP and knowledge discovery come into play. Furthermore, since Alexa needs to use external data sources to answer questions about (for example) the weather, some kind of data linking or querying is involved, including to sources such as Yelp (if the user is asking about restaurants and reviews) and Wikipedia (for general factoid questions).

Considering the significant investments made in Alexa (although the exact figure in dollar terms is not currently available, to our knowledge), and the functionalities listed above, there is little doubt that Alexa could not have been possible without both engineering commitments inside the company and AI advancements. Indeed, since at least 2019, Amazon has 90,000+ "skills" that users can download on Alexa-enabled devices. Only three years prior to that (2016), fewer than a thousand such skills had been available. Also, the AI Cortana service by Microsoft could be used on Alexa devices since 2018. More integration soon followed, including with the Wolfram Alpha answer engine (that allows users more accurate answers to questions on topical matters such as geography, math, and astronomy).

In contrast with Alexa, that much of the non-technical public is aware of due to the eponymous name of the devices that are used to access the virtual assistant, the Amazon Product Graph is a less visible effort for the general public. The Amazon Product Graph is an excellent example of a *knowledge graph* (KG). First popularized and brought to the public's attention due to the Google Knowledge Graph, and its impact on making search more seamless, KGs are structured representations of data that are more flexible than standard databases (such as Oracle and SQL Server) but less free-flowing and "open world" than natural language text. Designed correctly, this allows KGs to possibly achieve the best of both worlds: they can express information more intuitively and using natural language phrases, which can assist in flexible querying (as is common in search engines where ordinary users just type in keywords as opposed to proper queries that are common in structured fields such as medicine or finance) and uncovering of (previously) unknown relationships in large datasets using sophisticated graph mining techniques, while at the same time facilitating storage and access of the data in graph databases such as Amazon Neptune or Neo4j.

As we have shown extensively in our own prior research, building high-quality knowledge graphs is not only time-consuming but can involve the bridging of multiple fields of research [31, 33], including NLP, Semantic Web, data mining, standard machine learning, and (much more recently) more advanced graph representation learning [16,32,45,51,52,56]. Researchers in the field have been studying techniques for automating the different steps in knowledge graph construction and cleaning for at least a decade, with some steps of the pipeline having been studied in different fields of research,[4] and different application areas, over many decades (before being applied to KGs). An obvious example of such a step is information extraction, which is the NLP problem of extracting structured pieces of information (including "named entities" such as places, people, and organizations, but also relations such as "works for" between entity pairs), but also more advanced graph-theoretic techniques such as *entity resolution* [27, 28]. More recently, KGs have also been applied for social good, including crisis informatics [18, 36, 37, 43], and fighting human trafficking [20, 21, 24, 25, 35, 44, 50, 55].

In some domains, such as linking census records or movie information, there are not many relationships, and the text descriptions and labels tend to be very regular. Domain-specific techniques are also applicable [53, 54, 73, 79] and are needed for high quality, although the research community is now seeking to develop relatively domain-independent means of doing so as well [15, 29, 30, 41, 78]. For instance, there are not too many ways in which a person's name can be written (including using acronyms, short forms such as substituting "Jon" for "Jonathan," and so on), and when inferring whether two movies are identical, we can draw on the fact that numbers in the title are important (e.g., Mission Impossible 3 versus Mission Impossible 4), and should be accorded special weights.

These examples suggest that problems such as entity resolution can be customized to selected well-known domains, but it is much more difficult in a heterogeneous, high-variety domain such as retail products, where the types and numbers of relations between products can be practically "unbounded" and include not only physical attributes (e.g., color for rugs, and flavor for hummus), but also abstract attributes encoded within product text descriptions, news articles, public fora where people discuss the product, and customer reviews, to only name a few. In general, "matching" problems (whether of entities, textual descriptions, or even entire taxonomies) are difficult in such "diffuse" domains [26, 40, 42], both in terms of efficiency and effectiveness, as is rigorous manual annotation of true versus false matches [48, 49].

The retail domain is obviously Amazon's bread and butter, and the AI team in Amazon that works on the Amazon Product Graph has been able to significantly improve coverage and accuracy of the graph over the year. Their research effort has been documented in top machine learning and Artificial Intelligence conference. For example, at the premier Knowledge Discovery and Data Mining (KDD) conference in 2020, the researchers presented a system called AutoKnow, which comprises a collection of techniques that are able to augment product knowledge graphs automatically (leading to higher coverage of relevant information, but without much

---

[4] Discerning technical readers will recognize that this is just one of several graph-theoretic fields that are prominent in AI, where graphs have always played an important role. Within our own group, and others, these fields can often intersect and may also involve other disciplines. A particularly good example is *network science*, which is mature enough by now that several books have been written on it [58, 63]. Similar to KGs, network science can also be applied to many structured problems and domains [34, 38, 39, 47, 74].

loss of quality) with both structured and free-form text data sources. According to a leading researcher working on AutoKnow at the time, the system was able to increase the number of facts in the *consumables product graph* (which covers categories such as beauty, grocery, and baby) by almost 200%. Product types were identified with 87.7% accuracy, leading to negligible loss in quality of data at much higher coverage. Compared to the existing baseline systems, AutoKnow was able to demonstrate an improvement of more than 300% on specific KG construction problems, such as product type extraction (a domain-specific version of information extraction that was described earlier).

3.2.3   Meta

In 2016 alone, Facebook (which had not yet re-christened itself as Meta at that time) spent 21% of its sales revenue on R&D, amounting to nearly $5.9 billion spent on research. Within Big Tech, Facebook's R&D percentage is actually more of an outlier (i.e., significantly higher) than many people expect. In part, this may be because Facebook is one of the "newer" Big Tech companies and has faced consistently greater upheaval due to the emergence of video as an important modality for social media, and the negative criticism due to geopolitics (e.g., rise of bots and spread of misinformation). Although it is difficult to accurately estimate how much of this went to AI, industry sources firmly believe that it was certainly one of the top priorities. This was evidenced by its investments in its virtual assistant (called "M," perhaps in a sign of things to come), chatbots, augmented reality, news feed curation, and facial recognition.

Similar to some of the other Big Tech firms, Meta actively maintains an AI research page. In April 2022, for example, it announced a long-term research initiative to better understand how the human brain processes language, an endeavor that will likely take expertise in both AI (particularly deep learning-based language models) and neuroscience and is reminiscent of DeepMind's cross-disciplinary efforts, such as biology and AI. This is in addition to the existing efforts such as live face de-identification in videos, pose estimation, representation learning and computer vision efforts in e-commerce, video captioning, and fairness in machine learning. Similar to Google, their groups tend to publish quite extensively in top AI and machine learning conferences, and there are several instances of code and datasets being openly shared.

We note that open-source sharing and public release of code and selected datasets is an interesting trend that stands out among Big Tech companies today and is in contrast to a previous era. While not all models and datasets are released, and there is likely technology that is kept under wraps (or at least partially so), we are also seeing an unprecedented degree of openness from these research groups. This is a boon, not only for academic researchers and startups, but also non-Big Tech incumbents, to increase the pace of their innovation in AI and devise their own use cases without incurring the cost of building or training large-scale deep learning systems from scratch.

3.2.4   Other Big Tech: Microsoft and Apple

We round off our discussion on Big Tech with brief notes on AI efforts at Microsoft and Apple. Although we could provide a detailed description, the trends are sufficiently similar to those of the other three companies just described. Microsoft is well known for its AI research groups, similar to those at Meta and Google. A good example is the *Productivity and Intelligence* group, which (per the website) "brings together expertise from artificial intelligence, human–computer interaction, information retrieval, and software engineering" and has contributed its research and technology to improving well-known Microsoft products such as Office and Bing. Like some of the other Big Tech firms, these groups publish prolifically in established AI and machine learning conferences on topics as varied as recommendation systems, natural language processing, and AI for social good.

Apple has a reputation for being more secretive than the other Big Tech firms about its efforts and line of products (including rumors of developmental efforts ongoing in virtual and augmented reality), but its interest in AI can be evidenced by one of its best known acquisitions, Siri, in 2010. Originally, a spin-off from a project developed by the SRI International Artificial Intelligence Center, the voice assistant has since been integrated into Apple's line of products and uses voice queries, gesture-based control, focus-tracking, and a natural language user interface to answer questions, make recommendations, and perform actions by delegating requests to a set of Internet services.

3.2.5   Other Large Tech Firms in the United States

Although the companies we just discussed may definitively be considered Big Tech, there are other large technology companies that, while not necessarily "Big Tech," are big enough that they merit consideration here. In some cases, it is not the size of the company that makes it cohere less with the other Big Tech companies, but the focus on markets that have traditionally not been the sphere of Big Tech. One example is Tesla, which has an obvious focus on the electric vehicle market. Business Insider has claimed, for instance, that because of its emphasis on this market, Tesla should be classified more like Honda or the "Big Three" American automobile manufacturers based out of Michigan (General Motors, Ford, and Chrysler) [12].

However, the Tesla Autopilot system is a prototypical example of how AI can influence a staid industry like automobiles. Tesla itself describes this system as "an advanced driver assistance system that enhances safety and convenience behind the wheel." Currently, the technology is intended for use with an attentive driver at the wheel and should be thought of as an assistive technology. Indeed, a surprising number of modern industrial AI use cases fall within this umbrella, a phenomenon that is anecdotally accepted in real life, but which is disconnected from the *de facto* expectation within AI research to fully automate certain AI tasks. In Chap. 4, we discuss the phenomenon in more detail as *augmented AI*. As argued therein, including by drawing on sources from the consulting literature, augmented AI has ramifications for the workforce (mostly positive, but not equally for all), similar to how self-driving technology like the Tesla Autopilot can benefit drivers, but may eventually disenfranchise an entire class of workers (taxi and gig economy drivers) as it continues to evolve.

On a more digital front, Netflix and its recommendation algorithm are staples in business school case studies. Although we will not over-use the analogy, Netflix, similar to Tesla, is often cited as an example of how a new business model can upend a traditional market. In the case of Netflix, the original innovation was in offering *mail-order rental* when it was founded in 1997, and not the now ubiquitous streaming platform[5] that Gen-Z and younger millennials have come to associate it with.

Netflix's connection to AI is in its recommendation system. It was probably the first commercial enterprise to recommend entertainment-based content, such as movies, through such an algorithm. In the research community, recommendation has a rich history [9, 10], but arguably, the field never really took off until the mass adoption of the Web. According to the webpage of Netflix Research, recommendation algorithms (colloquially called "recommenders" by specialists) are at the "core" of Netflix's offerings. Recommendation is a challenging problem because there are many contextual factors (such as a user's mood, or even the day of the week, when they log in) that could potentially correlate with the recommendations that should be offered to them by the service. A major operational contribution of researchers and applied scientists in the information retrieval, machine learning, and other communities, which have published extensive research on recommenders, is based on the simple notion that two users that seem to be "similar" in their preferences could be used by a system to suggest good recommendations to each (an approach technically known as *collaborative filtering* [57]). For example, if an AI has determined that A is similar to B in that they both seem to like science-fiction movies from the 1970s, it may end up recommending a hit movie from the 1970s to A that A might not even be aware of. In the early days, Netflix's algorithm became famous for sometimes recommending movies that were not household names but ended up becoming favorites among a niche community (or even "sleeper" hits), owing to the recommender surfacing it to the "right" set of users.

Determining similarity in these multi-varied contexts can be addressed by machine learning techniques, which have become very advanced in recent years due to deep learning. Different notions of similarity may influence the quality of recommendations in ways that are not completely predictable. Deep learning can discover complex and non-linear patterns, but only if provided with enough data.

---

[5] In a sign of where the future was eventually headed, Netflix's initial foray into the subscription-based content model (1999) far predates the streaming, although subscriptions underpin the revenue of many streaming services today. The subscription model was discontinued only a few months after start, but was (obviously) brought back later.

Fortunately for companies such as Netflix, there is both an enormous customer base and internal company data (most important of which is who-watched-what, especially when they were presented with a list of recommendations) that can be used by these deep learning systems. This is the real reason why it is appropriate to consider Netflix as a technology company, rather than as a media company. Recently, of course, it has emerged as an important force in the latter due to its foray into original content production, some instances of which have been extremely successful (a paradigmatic, relatively recent example of which is "Squid Games"). However, it stands to reason that, as entertainment content explodes on all fronts (with even e-commerce giants such as Amazon now producing reams of original content), the winners of the "streaming wars" will not only be those who produce the best content, but also who are able to use AI and recommender systems to keep users coming back. The dominant success of Google in search is a prime example of how important information retrieval is, and recommenders are no different, especially when users are flooded with choices for what to watch next and where to spend subscription dollars in an increasingly inflationary and crowded market.

Although we considered Netflix and Tesla in this section, there are numerous other examples of large technology companies, including in the Bay Area, that are looking to significantly incorporate AI into their business. Examples that we could spend pages covering, but would largely make a similar point as the two examples earlier, include IBM, Cisco, HP, and several others. This larger point is that innovation and adoption in emerging technologies, of which AI is in an especially crucial phase due to both the maturity of the technology itself, and to underlying competitive market forces, are expected to be a cornerstone of growth for the companies that succeed in making a grab for the "money left on the table." The "money" is expected to be in the trillions over the decades to come.

3.2.6   The Chinese "Big Tech"

Although Silicon Valley gets flak for being a winner-takes-all environment, the investments in AI happening all across the board should make it clear that there is no clear "AI winner" in either the valley or anywhere else in the world. AI is not just a Big Tech phenomenon anymore, any more than it is a US-only phenomenon. This is in contrast with prior "technological revolutions," such as chip-making, where there was little interest from other industrial sectors, and to a large degree, other nations (except as manufacturers).

The emergence of the Chinese Big Tech or BATX (Baidu, Alibaba, Tencent, and Xiaomi) is a case in point of the difference between digital technologies and these earlier innovations. The first Chinese Big Tech companies started in the 2000s, coinciding with the explosive economic growth of China and its middle class. Today, the BATX firms are in good company. After 2015, other tech companies such as Huawei, DiDi, and ByteDance (which owns the TikTok service) have joined their ranks, albeit not to the same degree. There are direct parallels to the markets in which these companies operate and their US-based equivalents. We see these companies as servicing, in their markets, a global *demand* for certain ubiquitous technological needs, such as app-based ride-sharing and digital payment systems. However, China is unique compared to markets such as the European Union because of both its geopolitics and internal economic policies and trajectory, but also because of its language and culture. As such, although Western tech companies have attempted to make inroads into the market, and some (such as Apple) have succeeded to a degree, the uniqueness of the market and the ability of Chinese entrepreneurs to innovate in this space explain the rise of Chinese Big Tech as a phenomenon in its own right.

AI has increasingly come to be identified as an important component of these companies' growth over the last half-decade. An obvious example is recommending content on video-based social media (not dissimilar to Netflix, in principle), such as TikTok. The TikTok recommendation engine has become famous in recent years, owing to TikTok's uncanny ability to recommend videos that keep users coming back and staying longer. Although the algorithm is complex and not completely known, there is already a plethora of materials, mostly unproven, on the Web (not unlike SEO tips from an earlier era) on how to create videos that may end up at higher ranks on users' feeds. However, not everything is a secret. For example, TikTok has publicly shared the main outlines of its recommendation system [70]. Factors taken into account in the system include not only video data such as captions and hashtags, but also user engagement data, such as comments and likes. Some aspects of the algorithm are reportedly controversial. The Wall Street Journal reported through a video, how TikTok relies heavily on the *amount of time* that users spend watching the video [72], in order to recommend other videos to them that would keep them "scrolling." The algorithm can be addictive in its effects and, according to the report, could sometimes lead younger viewers to content that promotes self-harm or suicide. To their credit, however, TikTok has publicly said that it is working hard to delete content that violates its terms of service.

Other Chinese Big Tech are innovating on AI in similar ways, albeit in different domains. The key point to note is that these efforts mirror those of the Western counterparts in principle, but the details can be very different, due to the different characteristics of the Chinese language and market (both physical and online) compared to the West. Furthermore, because these companies are newer, they can learn from the mistakes of their Western counterparts rather than independently repeat them. One encouraging commonality is that several of these firms also have open-sourced parts of their code (on selected research projects) and datasets. An excellent example is Baidu, which open-sourced its deep learning platform PaddlePaddle, as well as its autonomous driving platform, Apollo. Therefore, it is safe to say that the trend toward open source, especially among Big Tech in both China and the USA, is global in scope, and a vital accelerant for developing domain-specific AI that can feed into the industries of the future.

3.3   Large Firms Outside Big Tech

More encouragingly, according to an event organized by Fortune in late 2021, not only is AI being increasingly seen as an important investment in large corporations, but they have also started to consider use cases for AI adoption beyond the "easy" or obvious ones (such as chatbots or robotic process automation). Some examples that were cited in a Fortune newsletter are [23]:

1. *Increasing personalization in personal finance by helping customers identify their individuals' needs:* As the fourth largest bank in the world, and an important player in the US retail banking market (with over 70 million customers), Wells Fargo needed to respond to growing consumer demands for a "frictionless" experience across the various services or "touchpoints" that it offered. The pandemic only intensified these needs, since consumers now have a greater preference for digital channels, including touchless experiences. In the modern digital economy, personalization remains an important driver of customer engagement, but this can be challenging for any company (especially one in an industry that has not historically been associated with personalization or recommender systems).
   Wells Fargo addressed these challenges through its Pega's Customer Decision Hub, which was designed using real-time modeling and adaptive machine learning. Pega can be used by the bank to dynamically determine an individual's "next best conversation" even as those individuals are engaged in an interaction [65]. Pega is used to deliver, on a continuous basis, next-based conversations not only on the mobile app, but also the retail branch, outbound channels, or contact center that the customer may be engaged with or in at that time. The stated goal was to help increase relevance of messages delivered to customers through these channels and also to introduce "new, pandemic-influenced" conversations that would enable customers who were struggling to become more financially resilient.
   The effort seems to have paid off, since not only were customer engagement rates found to have increased by up to an order of magnitude, depending on the modality (or channel) and specific use cases being considered, but conversion rates noticeably increased across these channels. According to then Head of Personalization at Wells Fargo, Giles Richardson, the services helped the company to deliver "personalized conversations at true enterprise scale—spotting patterns from billions of interactions so the customer gets the right message and the best experience" [65].
2. *Automatically analyzing video data collected by drones flying over solar panel farms to proactively identify problems:* According to a 2021 article in VentureBeat [60], Duke Energy used computer vision and robots to cut costs by $74M. This was not an idle investment, but likely an important strategic response to challenges that the company was facing, including pressure to transition to clean energy with the goal of reaching net zero emissions by mid-century. As an essential service, Duke Energy is responsible for supplying the daily electricity

needs of almost 25 million residents, and as a utility company, its culture is built around ensuring reliable and safe services.

On the surface, it may not seem as if AI can do much in this transition, but that would be misleading. As Duke Energy's then chief information officer explained, inspecting solar panels and farms can be time-consuming, risky, and labor-intensive. For instance, a single unit can take over 40 h to inspect, with a "regular solar site" having anywhere from 20–25 units that would need to be inspected. An additional, and more important, concern is that the inspection task can also be dangerous. Solar sites can span hundreds of acres, and technicians have to conduct the inspection with heat guns, not to mention touching live wires. To make this task easier, more streamlined, and less risky, Duke Energy reportedly started experimenting with drone technology outfitted with infrared cameras. This way, technicians could depend upon the images taken by the drones as a first step, which would guide them to the sites that are experiencing faults and issues (or likely to). The task may seem easy but in fact requires "stitching" together thousands of images using advanced computer vision algorithms. The net result of using such technology is both increased safety and efficiency. Furthermore, although we use the documented example of Duke Energy here, it is likely that as the technology becomes easier to use and takes less time, energy, money, and consultancy to set up and deploy, utility and clean energy companies all over the world may start to rapidly adopt it as an essential productivity-enhancing technology .

3. *Optimizing airplane performance and airport construction:* AI is often associated with software and abstraction (or on the other extreme, with advanced robotics), but it can have a real impact on other high-tech industries, such as aviation, that involve many complex hardware and software components. A case study is Boeing, which in late 2020 completed a sequence of test flights that were designed to explore how high-performance aircraft without crews could be jointly operated and controlled by AI using a combination of onboard commands and data sharing. In this exploration, Boeing added aircraft one at a time over a period of 10 days until five such aircraft managed to operate as a single autonomous "unit" reaching speeds of over 160 miles/hour [3].

Emily Hughes, then the director of Phantom Works (Boeing's prototyping arm for its defense branch), was quoted in a statement as saying that the tests demonstrated the success of AI algorithms to "teach" the "aircraft's brain" the requirements that it needed to understand [3]. While the test flights were evidently part of Boeing's defense-related business, the company has also gone on record to state that technologies developed as a consequence of this program could potentially apply to all of its future autonomous aircraft. Additionally, Boeing's subsidiary (Aurora Flight Sciences) has also been reportedly constructing autonomous flight vehicles that are smaller and that would likely be more useful in nimbler operations, e.g., urban mobility aka "flying taxis" [71].

What about Boeing's storied rival, Airbus ? Considering the historic and present rivalries between these two companies, it is not surprising that Airbus has also been exploring the use of AI. In early 2020, it was able to conclude the first fully automated take-off and landing based on computer vision. This feat fell within the scope of its *Autonomous Taxi, Take-Off and Landing* project. Typically, the take-off relies on an instrument landing system, but in this case, it could be managed by image recognition software in the aircraft. Although the project itself was completed mid-2021, Airbus has gone on record to state that autonomous flight is not their goal in itself (in other words, pilots remain central to operations); rather, they want to use similar such technology and its complements to improve flight performance, not dissimilar to Boeing's use case.

4. *Predicting loads and usage patterns in a major city's subway system using image recognition technology:* In our last use case outside of Big Tech, we note Accenture's efforts in helping the Metro de Madrid (the Madrid subway system) implement an AI-driven ventilation system that is adaptive and aims to minimize both energy costs and emissions (hence ensuring higher air quality in the metro stations, as well as increasing the comfort of commuters) [1]. The implemented system was documented as having achieved significant impacts prior to the pandemic: it reduced ventilation energy costs by almost a quarter and eliminated annual CO_2 emissions by almost 1,800 tons. Reportedly, the system also includes a simulation engine (and maintenance module) that allows technical staff and planners to track certain operational failures. All of these technologies help Madrid's subway not only to track and manage energy consumption, but also to proactively identify the deficiencies of the system and conduct maintenance of the equipment.

3.4 Startups and Small/Medium-Sized Enterprises (SBEs)

It is not uncommon today to see news that rings the alarm bell on the growing quasi-monopolistic power of the Big Tech firms. However, Big Tech should not be singled out in this regard. Mergers and acquisitions were at an all-time high in recent years, likely fueled by low interest rates and a large appetite for risk. To make matters more grim, small- and medium-sized businesses were pummelled in the wake of COVID-19. Startups have fared better, due to investments by venture capitalists and private equity, and increasing valuations of such firms in "hot" sectors such as blockchain, cybersecurity, virtual reality, and AI. Technically, these can be considered small enterprises, although in popular parlance, a "small business" is taken to mean something different from a "startup."

Terminology aside, there is hope that smaller companies may hold an advantage if they start out as "digital natives" from the get-go. This is not just due to investor appetite or preference, but due to the advent of cloud computing and other technologies that make it increasingly easier for nimble companies to set up complex infrastructure without too much fixed cost. In the long run, costs would seem to favor in-house data and compute infrastructure rather than on-demand cloud computing. In the short run, the variable cost structure is precisely what allows many startups to get started in the first place, especially if they need to get a working prototype ready to entice the next round of investors.

More recently, there has been much excitement about "language" AI startups. This is widely credited due to the rise of transformer-based deep neural networks. We had mentioned these networks earlier in the context of Google incorporating the BERT language model into its search engine, but subsequently, we provide even more details on the potentially groundbreaking nature of this technology. The impact of this technology on the startup space can be summed up in the title of a recent Forbes article that states that a "wave of billion-dollar language AI" startups is just around the corner [75]. They liken the current era to an "inflection" point where research in this space is amenable to rapid real-world adoption without the decades-long friction that often accompanies the transition of fundamental research to actual applications.

Examples cited in the article include Hugging Face (extremely popular in the language AI community), Cohere, Primer, AI21 Labs, and OpenAI. Not all are based in the United States. AI21 Labs, for instance, is based out of Israel and offers proprietary (and often, large) language models to customers via Application Programming Interfaces (APIs). These APIs can be used to integrate language model services with customer applications. Nor are the startups solely focused on traditional consumers or businesses. Primer, which was started before the trans-former was even invented, has clients in spaces such as defense and government.

At the same time, in some sectors of the economy, automation is being explored in areas that have *not* been widely covered in the press. Earlier, we had mentioned how Boeing and Airbus have been investing in AI. There are also startups investing in the space of automated aviation. According to the headline in [5], pilot-less planes may be around the corner and will likely be hydrogen-powered. For example, the UK government itself is funding, to the tune of almost $2.5M, a novel project called HEART (Hydrogen Electric and Automated Regional Transportation) that seeks to develop a regional air transport network that will see passengers flying on hydrogen-powered (and pilot-less) aircraft for short regional hops by the middle of this decade. Other automation aviation startups include Joby Aviation, Relativity Space, and SpaceX, which currently stands as the leader in the space and is even considered to be the world's most valuable startup.

Interestingly, these examples and the sectors they are operating in provide some inkling of *what* industries of the future may look like. Although we noted at the outset of the book that our primary purpose herein was not to engage in technological prediction (which history has shown is unwise, as we are prone to both underestimate and overestimate innovations and their impacts), the startups that are growing their market shares, and acquiring copious amounts of funding, provide the best guide to such prediction. Transformer-based neural networks and language models, into which we delve in the next section as a mini-case study of a specific AI technology having tremendous impact, seem likely to create a generation of unicorns in domains ranging from legal tech and EdTech to greater automation in the software engineering industry itself (as we briefly discuss in the case study). Aviation and space are also important new frontiers where AI will continue to play a supporting role. E-commerce, virtual reality (the "metaverse"), and cybersecurity are examples of applications or environments where the main product does not necessarily revolve around AI, but where AI may end up providing a competitive advantage as the market becomes increasingly saturated. AI will also play a role in the clean energy industry, including predictive maintenance and twenty-first century infrastructure endeavors such as the construction and maintenance of "smart grids." In agricultural biotech and advanced manufacturing, hardware innovations (such as cheaper and more reliable drone technology, and edge computing IoT devices) are intersecting with AI in interesting ways.

We end with a note on important startup economies outside of the United States and Western Europe. China, India, and Israel have all emerged as important incubators of startups, with Israel often compared to a nascent Silicon Valley. A 2021 Forbes article, for instance, was titled "Is Israel The Next Silicon Valley?" [6]. The funding statistics alone paint a telling picture. Tech startups in Israel raised almost $18 billion in the first three quarters of 2021, which was almost double the total amount of funding received by Israeli startups in 2020. Quite a few of these were "mega-rounds" that involved deals over $100 million. Several AI startups have been beneficiaries of this funding, although AI is not the only emerging tech experiencing high growth in that region. Examples include Versatile, a construction-tech startup using AI for optimization of construction (and other related) processes, and Sisense, a business intelligence software company that has 2,000+ customers globally.

3.5   Case Study: Neural Language Models

Much of the discussion thus far has been on the companies, and their market presence, positioning and future potential, but we now switch to an example AI technology that will almost definitely play an increasingly important role in applications developed in organizations from all of the different groups just covered (startups and SMEs, large non-Big Tech corporations, and Big Tech and their Chinese equivalent). In fact, such adoption is already underway, and picking up steam. The technology we are referring to here is that of *neural language models*, which are now assumed within the community to mean several "generations" of language models that are based on a specific kind of neural network called a *transformer*. Note that a generation here usually only lasts several months, maybe a year, in a testament to the extremely rapid pace at which new models are being developed, trained, and released.

Why did we pick neural transformer-based language models (henceforth just called language models) as our case study of choice? Perhaps the best reason is their performance, which has been startlingly human-like. Language models have now been shown to be able to process and manipulate language much more robustly and context-dependently than previous systems, and they have achieved state-of-the-art performance across a range of NLP tasks, including machine translation, question answering, text generation, and long-text summarization. Beyond performance, there is another important technological element that may have been largely responsible for the widespread adoption of the language models in both industry and academia. Usually, when a model is designed for a task (such as machine translation), it has to be specially trained for it and requires a large body of linguistic resources and training data. However, the most common language models today are first *pre-trained* (usually at significant cost, by industrial organizations) on a large set of books, documents, and other natural language text collected from the Internet (and occasionally, proprietary sources). Once pre-trained, the model can now be customized or *fine-tuned* at much lower cost (even just using a single desktop) for specific tasks. For example, a pre-trained model could be fine-tuned, using training data, for question answering, and the same pre-trained model could also (independently) be fine-tuned for information extraction, yielding two different models that share the same "base." In a manner of speaking, the difficulty and cost of pre-training get effectively amortized across many applications and domains.

The concept of transformers was first introduced within Big Tech, specifically, at the Google Brain research lab. These neural networks are especially adept at processing long sentences (and more generally, *sequences*) and uncovering relationships between phrases and words much more accurately and comprehensively than the previous generation of NLP techniques (which itself depended on a different type of deep learning). Progress since that first system has been rapid: today, almost all of the Big Tech firms are known to be using, or otherwise developing their own, language models. The research company OpenAI, which has been supported (at various times) by the likes of Microsoft and Elon Musk, is famously known for the GPT-3 language model [7], which is capable of *generating* natural language text, given simple input prompts, that can be difficult to distinguish from human-generated text.

Would it be safe to declare NLP problems to have been "solved" by these language models? For better or for worse, there is still a long way to go. One of the concerns that has become especially loud given increasing awareness of climate change and global warming is that progress through these models is being achieved by making them ever larger. A recent model developed by Google, called Pathways Language Model (PaLM), uses more than half a trillion variables at a time. At this pace, it will not be long, according to industry watchers, before trillion-parameter language models become common. The reason why this is relevant to the conversation on climate change is that training such large models takes enormous energy and resources. As noted earlier, although language models generally only have to be pre-trained once or twice (since they can then be "fine-tuned" for specific applications and domains), even just that one-time pre-training can cost many millions of dollars and consume the same amount of energy that several homes (at minimum) in the United States consume during an entire *year*. The criticism here is similar to that lobbied at mining a cryptocurrency such as Bitcoin [77], the mining of which can take the same energy consumption as that of a small country.

3.5.1   Can Transformers Automate Software Engineers?

Software engineers have continued to be high in demand in the workforce, especially over the last two decades. This demand is often cited as a need for more liberal immigration policies [61], and many companies have even considered hiring students who do not have college degrees (or a college degree in computer science) but may instead have been self-taught, or graduates of coding "bootcamps" [8]. In contrast, for more advanced research-based positions, including in industry labs, a PhD is the norm rather than the exception.

Because of the six-figure salaries that software engineers command, and the relatively rapid turnover in skills (with languages and software packages falling in and out of fashion with the decade), it is natural to expect companies to invest more in enhancing productivity in this area. This does not necessarily involve AI. Good examples include Integrated Development Environments or IDEs, which allow efficient writing, organization, and testing of code. An IDE such as PyCharm [22], for instance, enables programmers to quickly and easily navigate code and run unit tests in a visually appealing and intuitive manner, not to mention generate documentation, download packages and dependencies, facilitate version control (including uploading and fetching code from GitHub from within the IDE itself), and make use of many other productivity-enhancing facilities for writing, publishing, and managing complex code.

While this is already major progress compared to simple text-based editors (although competent programmers still insist on their utility, especially when only a command line is available, as is common when deploying code on the cloud or on servers), and has arguably democratized software engineering for a younger generation of computer scientists, the code can still be overwhelming and overly complicated. At its core, like many modern knowledge-intensive tasks, code-writing and other related software engineering tasks are *abstract* and, as any software engineer will attest to, can be surprisingly brittle at times. Seemingly obvious mistakes, for example, where the intent of the programmer was clear, but a minor error of logic may have inadvertently occurred, could end up crashing entire systems and lead to hand-wringing days of debugging.

Transformers and other neural network technologies are now at a point where they can start to serve as active assistants in this process, leading to less stress and more productivity from programmers and their employers alike. This is not speculation or even industry hype: a recent paper published in a premier international machine learning conference went deep into how a neural network could start to understand programs [64]. Since then, news has also emerged that DeepMind is working on a similar problem, an effort called AlphaCode [59]. Although it is not yet evident that AlphaCode can fully and consistently outperform star programmers (who are largely hired by the Big Tech companies), it does show promise already, even at competition-level problems, and given DeepMind's success in other complex arenas where humans long outperformed machines, we suspect the day is not far when later versions of AlphaCode will prove to be formidable competition for the current software engineers. The beneficiary of this technology will obviously be Google, but others may soon follow, since the techniques for accomplishing this task are well understood by expert researchers (although by no means, any easier to cost-effectively implement at the time of writing).

3.5.2   Applications Beyond NLP

Transformer neural networks are powerful enough that they are already finding uses outside NLP, even in the research community. An excellent case study is computer vision. Researchers have demonstrated that, on several important visual benchmarks, transformer-based models perform nearly at the same level as other types of (now more traditional) approaches such as convolutional and recurrent neural networks. However, given that transformers may be less prone to an inductive bias that is vision-specific, and also because of their state-of-the-art ability to incorporate natural language inputs into vision models (making them particularly good fits for multi-modal reasoning), they are receiving increasing attention from the computer vision community. For the interested technical reader, we point to a relatively recent review of "vision transformer" models [19]. In the review, the authors categorize the models by the different tasks they seem well-suited for and enumerate their advantages and disadvantages. The categories that they review include the backbone network, video processing, as well as low-level and high/mid-level vision. The review also considers efficient methods for enabling execution of transformers on real devices, which may be an important driver of an industry such as advanced manufacturing, IoT, and edge computing.

While all of this is promising, we note that many challenges still remain, and further research is required, as the review is careful to point out. However, given the rapid pace of innovation that is happening in this area in academia and industry alike, we believe that it is only a matter of time before transformers become a staple in practical and industrial applications that need to make joint use of multi-modal streams of data (such as images and video feed) and text data.

3.5.3   Potential Ethical Concerns

Similar to deep learning models used in computer vision, ethical concerns have also been raised about some of these large-scale language models. Earlier, we stated how expensive it can be (from both cost and energy perspectives) to train these models. The rising popularity and adoption of these models has raised many concerns about how their carbon footprint will contribute to global warming and climate change. Unfortunately, climate change is not the only ethical issue with these models. A growing body of scholarly work has shown that these algorithms can incorporate human biases from their massive pre-training corpora, involving sensitive and important issues that we continue to grapple with in the public and corporate sphere today, such as race, religion, sexual orientation, and gender [2,62].

The models cannot only be coaxed to use toxic and offensive language, but they may also have worse performance on data points involving under-represented groups.

A promising sign is that discussion and potential mitigation of such biases are now being recognized as an increasingly serious problem in the AI community itself. Researchers are therefore incentivized to improve the models and flag the concerns much earlier.

3.5.4   Summary

This section went (a little) deeper into a specific technology that has recently taken AI by storm: transformer-based neural networks. These networks, also called neural language representation models or just language models, when applied in the context of modern NLP research, have proven to achieve near-human accuracy on a variety of difficult tasks. Examples include speech recognition, machine translation, question answering and chatbots, and even "general" AI problems such as commonsense reasoning [11,67–69,76], and [46]. We also briefly discussed work showing that they could be applied beyond NLP, including to other domains such as computer vision.

Given their breadth of application, and their replicated high performance across many independent researchers and teams, transformers have proven to be at least as influential (if not more, from a practical standpoint) as the deep reinforcement learning systems built by companies such as DeepMind that have bested humans in ancient games such as Go and Shogi. It is likely that they will play an important role in industries of the future, and we are already starting to see some applications and startups emerge in this space. A specific such industry might be automatic coding, which we believe is an under-appreciated market sector that will be revolutionized by such technologies. There has already been a shift toward writing code more easily (or with a shorter learning curve) for at least a decade now, and the movement toward "no-code programming" (such as by using visual interfaces) has also been experiencing a resurgence. Transformers' application to automatic programming is aligned with this movement, and one of obvious incentive to companies, including Big Tech. It is less certain how this might influence the workforce. In the next chapter, we dive deeper into the complexity of augmented AI and its potential impact on the workforce of the future.

There is much more that we can say about both the techniques and applications behind language models and transformers, but that is beyond the scope of this book. For example, one area that we did not touch upon was *creative* text generation, such as fictional writing and poetry. The transformer-based GPT-3 *generative* model has been found to be surprisingly adept at writing (among other things) poetry and even movie scripts [7]. Commentators have suggested that the quality is improving, although replacement of human writers does not currently seem to be on the horizon.

However, it is important to note that transformers have not been around for a decade at the time of writing. Therefore, it is not a matter of science fiction to hypothesize that by the year 2030, we may be watching a movie the script of which was largely written by a descendant of GPT-3. Regardless of how specific industries will be affected, it is certainly not speculative (given the current evidence on both performance and adoption of these models) to suggest that the occurrence of such effects is more a question of "when" rather than "if."

3.6   Conclusion

We close this chapter with a note on the role that government has ostensibly played in this evolution of this technology and its greater adoption in industry. On the regulatory or legislative front, to our knowledge, no one is seriously advocating for a purely *laissez faire* approach to the problem of regulating AI, in part due to the monopolistic-like tendencies of Big Tech, and the black box, potentially discriminatory nature of AI (which has been borne out by several studies in the peer-reviewed literature). There even seems to be rising acknowledgement that the role and definition of anti-trust itself may have to be re-defined, with a less overwhelming focus on consumer prices and more focus on what it means in the real economy to have one or two large companies dominating a specific sector, even if it does not necessarily lead to higher prices (at least in dollar terms) for consumers. This is a complex issue, with arguments that can favor either side.

As a consumer, the government (and especially, defense) can be important. We mentioned cases of startups earlier where the government is a primary consumer. Another noteworthy example, although not a startup, is Palantir Technologies (which is publicly listed on the New York Stock Exchange and has a market cap of over $20B as of May 2022). Although it now counts private companies also as clients, its original (and current) clients were federal agencies of the United States Intelligence Community. From its founding, the company's aim was quoted to be the reducing of "terrorism while preserving civil liberties" [17]. Contracts with the government can also be lucrative for the cloud computing arms of Big Tech giants. A case in point is the contentious (and now called-off) Joint Enterprise Defense Infrastructure or JEDI deal. This deal was going to be a $10B cloud contract that became the subject of a legal tussle between Microsoft and Amazon. Originally, in 2019, Microsoft was awarded the contract, which was intended by the Pentagon to modernize its IT operations and would have lasted as long as 10 years. However, Amazon's cloud computing unit, Amazon Web Services, filed a lawsuit protesting the decision on an alleged basis (in the lawsuit) of bias and undue influence. Perhaps due to the ensuing controversies and legal headache, the Pentagon has canceled the cloud contract and announced in a 2021 press release that it was launching a new "multivendor" cloud computing contract called the Joint Warfighter Cloud Capability, which is planned for December 2022 [14]. We do not believe such contracts, or calls for contracts, are exceptions. As more countries modernize their digital and AI infrastructure, especially in the face of new threats and policy issues, more such opportunities will arise, and both startups and Big Tech will be primed to take advantage.

Last, but not least, the government has long been an incubator of facilitating transitioning of academic research into companies, patents, and products. It is often not recognized fairly for its role in facilitating this transition in the popular press. Excellent examples of such an initiative that is known to, and used by, many startup and small business founders are the US National Science Foundation's *Small Business Innovation Research (SBIR)* and *Small Business Technology Transfer (STTR)*, which are both part of what is popularly known as the *America's Seed Fund* [13]. The fund is focused on "transforming scientific and engineering discoveries into products and services with commercial and societal impact." Typically, these discoveries have been made through scientific research and (therefore) have scientific validity, but their commercial success has not been validated. Each year, awards totaling more than two hundred million dollars (and spread across 400+ startups in the USA) are made to boost in R&D funding. The total dollars amount per startup is usually small compared to Series rounds (hence the fund being aptly named a "seed" fund), but a strong advantage is that the fund does not take equity in exchange for funding. The program is currently housed within the Directorate for Technology, Innovation and Partnerships in the NSF.

References

1. Accenture helps Metro de Madrid balance energy efficiency and passenger comfort with AI-based self-learning ventilation system (2019). URL https://newsroom.accenture.com/news/accenture-helps-metro-de-madrid-balance-energy-efficiency-and-passenger-comfort-with-ai-based-self-learning-ventilation-system.htm
2. Abid, A., Farooqi, M., Zou, J.: Persistent anti-Muslim bias in large language models. In: Proceedings of the 2021 AAAI/ACM Conference on AI, Ethics, and Society, pp. 298–306 (2021)
3. Ahlgren, L.: How airbus and Boeing are using artificial intelligence to advance autonomous flight? (2021). URL https://simpleflying.com/airbus-boeing-artificial-intelligence-flight/
4. Allyn, B.: Amazon's Alexa could soon speak in a dead relative's voice, making some feel uneasy (2022). URL https://www.npr.org/2022/06/23/1107079194/amazon-alexa-dead-relatives-voice
5. Bailey, J.: Pilotless, hydrogen-powered planes could be in the UK within a decade (2021). URL https://simpleflying.com/pilotless-hydrogen-regional-planes/
6. Bino, E.: Is Israel the next silicon valley? (2021). URL https://www.forbes.com/sites/eyalbino/2021/10/21/is-israel-the-next-silicon-valley/?sh=26155009177f
7. Brown, T.B., Mann, B., Ryder, N., Subbiah, M., Kaplan, J., Dhariwal, P., Neelakantan, A., Shyam, P., Sastry, G., Askell, A., Agarwal, S., Herbert-Voss, A., Krueger, G., Henighan, T., Child, R., Ramesh, A., Ziegler, D.M., Wu, J., Winter, C., Hesse, C., Chen, M., Sigler, E., Litwin, M., Gray, S., Chess, B., Clark, J., Berner, C., McCandlish, S., Radford, A., Sutskever, I., Amodei, D.: Language models are few-shot learners. CoRR **abs/2005.14165** (2020). URL https://arxiv.org/abs/2005.14165
8. Burke, Q., Bailey, C., Lyon, L.A., Green, E.: Understanding the software development industry's perspective on coding boot camps versus traditional 4-year colleges. In: Proceedings of the 49th ACM Technical Symposium on Computer Science Education, pp. 503–508 (2018)
9. Das, D., Sahoo, L., Datta, S.: A survey on recommendation system. International Journal of Computer Applications **160**(7) (2017)

10. Dau, A., Salim, N.: Recommendation system based on deep learning methods: a systematic review and new directions. Artificial Intelligence Review **53**(4), 2709–2748 (2020)
11. Davis, E., Marcus, G.: Commonsense reasoning and commonsense knowledge in artificial intelligence. Communications of the ACM **58**(9), 92–103 (2015)
12. DeBord, M.: Everyone who thinks tesla is a tech company is completely wrong – tesla should aspire to be Honda. (2021). URL https://www.businessinsider.com/why-tesla-is-not-a-tech-company-2019-2
13. Foundation, N.S.: About America's Seed Fund powered by NSF (2022). URL https://seedfund.nsf.gov/about/
14. Garamone, J.: Joint warfighting cloud capability award planned for December. (2022). URL https://www.defense.gov/News/News-Stories/Article/Article/2984496/joint-warfighting-cloud-capability-award-planned-for-december/
15. Getoor, L., Machanavajjhala, A.: Entity resolution: theory, practice & open challenges. Proceedings of the VLDB Endowment **5**(12), 2018–2019 (2012)
16. Gheini, M., Kejriwal, M.: Unsupervised product entity resolution using graph representation learning. In: J. Degenhardt, S. Kallumadi, U. Porwal, A. Trotman (eds.) Proceedings of the SIGIR 2019 Workshop on eCommerce, co-located with the 42nd International ACM SIGIR Conference on Research and Development in Information Retrieval, eCom@SIGIR 2019, Paris, France, July 25, 2019, *CEUR Workshop Proceedings*, vol. 2410. CEUR-WS.org (2019). URL http://ceur-ws.org/Vol-2410/paper26.pdf
17. Greenberg, A.: How a 'deviant' philosopher built Palantir, a CIA-funded data-mining juggernaut (2013). URL https://www.forbes.com/sites/andygreenberg/2013/08/14/agent-of-intelligence-how-a-deviant-philosopher-built-palantir-a-cia-funded-data-mining-juggernaut/?sh=4ed155947785
18. Gu, Y., Kejriwal, M.: Unsupervised hashtag retrieval and visualization for crisis informatics. CoRR **abs/1801.05906** (2018). URL http://arxiv.org/abs/1801.05906
19. Han, K., Wang, Y., Chen, H., Chen, X., Guo, J., Liu, Z., Tang, Y., Xiao, A., Xu, C., Xu, Y., et al.: A survey on vision transformer. IEEE Transactions on Pattern Analysis and Machine Intelligence (2022)
20. Hundman, K., Gowda, T., Kejriwal, M., Boecking, B.: Always lurking: Understanding and mitigating bias in online human trafficking detection. CoRR **abs/1712.00846** (2017). URL http://arxiv.org/abs/1712.00846
21. Hundman, K., Gowda, T., Kejriwal, M., Boecking, B.: Always lurking: Understanding and mitigating bias in online human trafficking detection. In: J. Furman, G.E. Marchant, H. Price, F. Rossi (eds.) Proceedings of the 2018 AAAI/ACM Conference on AI, Ethics, and Society, AIES 2018, New Orleans, LA, USA, February 02–03, 2018, pp. 137–143. ACM (2018). DOI URL https://doi.org/10.1145/3278721.3278782
22. Islam, Q.N.: Mastering PyCharm. Packt Publishing Ltd (2015)
23. Kahn, J.: Lessons from A.I.S. rare pandemic success (2021)
24. Kapoor, R., Kejriwal, M., Szekely, P.A.: Using contexts and constraints for improved geo-tagging of human trafficking webpages. In: P. Bouros, M. Sarwat (eds.) Proceedings of the Fourth International ACM Workshop on Managing and Mining Enriched Geo-Spatial Data, Chicago, IL, USA, May 14, 2017, pp. 3:1–3:6. ACM (2017). DOI URL https://doi.org/10.1145/3080546.3080547
25. Kapoor, R., Kejriwal, M., Szekely, P.A.: Using contexts and constraints for improved geotagging of human trafficking webpages. CoRR **abs/1704.05569** (2017). URL http://arxiv.org/abs/1704.05569
26. Kejriwal, M.: Disjunctive normal form schemes for heterogeneous attributed graphs. CoRR **abs/1605.00686** (2016). URL http://arxiv.org/abs/1605.00686
27. Kejriwal, M.: Populating a linked data entity name system. AI Matters **3**(2), 22–23 (2017). DOI URL https://doi.org/10.1145/3098888.3098897
28. Kejriwal, M.: Populating a Linked Data Entity Name System - A Big Data Solution to Unsupervised Instance Matching, *Studies on the Semantic Web*, vol. 27. IOS Press (2017). DOI URL https://doi.org/10.3233/978-1-61499-692-7-i

29. Kejriwal, M.: Predicting role relevance with minimal domain expertise in a financial domain. CoRR **abs/1704.05571** (2017). URL http://arxiv.org/abs/1704.05571
30. Kejriwal, M.: Predicting role relevance with minimal domain expertise in a financial domain. In: Proceedings of the 3rd International Workshop on Data Science for Macro-Modeling with Financial and Economic Datasets, DSMM@SIGMOD 2017, Chicago, IL, USA, May 14, 2017, pp. 10:1–10:2. ACM (2017). DOI URL https://doi.org/10.1145/3077240.3077249
31. Kejriwal, M.: Domain-specific knowledge graph construction. Springer (2019)
32. Kejriwal, M.: Unsupervised DNF blocking for efficient linking of knowledge graphs and tables. Inf. **12**(3), 134 (2021). DOI URL https://doi.org/10.3390/info12030134
33. Kejriwal, M.: Knowledge graphs: A practical review of the research landscape. Inf. **13**(4), 161 (2022). DOI URL https://doi.org/10.3390/info13040161
34. Kejriwal, M., Dang, A.: Structural studies of the global networks exposed in the Panama papers. Appl. Netw. Sci. **5**(1), 63 (2020). DOI URL https://doi.org/10.1007/s41109-020-00313-y
35. Kejriwal, M., Ding, J., Shao, R., Kumar, A., Szekely, P.A.: Flagit: A system for minimally supervised human trafficking indicator mining. CoRR **abs/1712.03086** (2017). URL http://arxiv.org/abs/1712.03086
36. Kejriwal, M., Gilley, D., Szekely, P.A., Crisman, J.: THOR: text-enabled analytics for humanitarian operations. In: P. Champin, F. Gandon, M. Lalmas, P.G. Ipeirotis (eds.) Companion of the Web Conference 2018, WWW 2018, Lyon, France, April 23–27, 2018, pp. 147–150. ACM (2018). DOI URL https://doi.org/10.1145/3184558.3186965
37. Kejriwal, M., Gu, Y.: A pipeline for post-crisis Twitter data acquisition. CoRR **abs/1801.05881** (2018). URL http://arxiv.org/abs/1801.05881
38. Kejriwal, M., Gu, Y.: Network-theoretic modeling of complex activity using UK online sex advertisements. Appl. Netw. Sci. **5**(1), 30 (2020). DOI URL https://doi.org/10.1007/s41109-020-00275-1
39. Kejriwal, M., Kapoor, R.: Network-theoretic information extraction quality assessment in the human trafficking domain. Appl. Netw. Sci. **4**(1), 44:1–44:26 (2019). DOI URL https://doi.org/10.1007/s41109-019-0154-z
40. Kejriwal, M., Miranker, D.P.: Experience: Type alignment on DBpedia and Freebase. CoRR **abs/1608.04442** (2016). URL http://arxiv.org/abs/1608.04442
41. Kejriwal, M., Miranker, D.P.: Local, domain-independent heuristics for the FEIII challenge: Lessons and observations. In: Proceedings of the Second International Workshop on Data Science for Macro-Modeling, DSMM@SIGMOD 2016, San Francisco, CA, USA, June 26–July 1, 2016, pp. 17:1–17:2. ACM (2016). DOI URL https://doi.org/10.1145/2951894.2951911
42. Kejriwal, M., Miranker, D.P.: Self-contained NoSQL resources for cross-domain RDF. CoRR **abs/1608.04437** (2016). URL http://arxiv.org/abs/1608.04437
43. Kejriwal, M., Peng, J., Zhang, H., Szekely, P.A.: Structured event entity resolution in humanitarian domains. In: D. Vrandecic, K. Bontcheva, M.C. Suárez-Figueroa, V. Presutti, I. Celino, M. Sabou, L. Kaffee, E. Simperl (eds.) The Semantic Web - ISWC 2018 - 17th International Semantic Web Conference, Monterey, CA, USA, October 8–12, 2018, Proceedings, Part I, *Lecture Notes in Computer Science*, vol. 11136, pp. 233–249. Springer (2018). DOI URL https://doi.org/10.1007/978-3-030-00671-6_14
44. Kejriwal, M., Schellenberg, T., Szekely, P.A.: A semantic search engine for investigating human trafficking. In: N. Nikitina, D. Song, A. Fokoue, P. Haase (eds.) Proceedings of the ISWC 2017 Posters & Demonstrations and Industry Tracks co-located with 16th International Semantic Web Conference (ISWC 2017), Vienna, Austria, October 23rd - to - 25th, 2017, *CEUR Workshop Proceedings*, vol. 1963. CEUR-WS.org (2017). URL http://ceur-ws.org/Vol-1963/paper613.pdf
45. Kejriwal, M., Sequeda, J.F., Lopez, V.: Knowledge graphs: Construction, management and querying. Semantic Web **10**(6), 961–962 (2019). DOI URL https://doi.org/10.3233/SW-190370

46. Kejriwal, M., Shen, K.: Do fine-tuned commonsense language models really generalize? CoRR **abs/2011.09159** (2020). URL https://arxiv.org/abs/2011.09159

47. Kejriwal, M., Shen, K.: Can scale-free network growth with triad formation capture simplicial complex distributions in real communication networks? CoRR **abs/2203.06491** (2022). DOI URL https://doi.org/10.48550/arXiv.2203.06491

48. Kejriwal, M., Shen, K., Ni, C., Torzec, N.: An evaluation and annotation methodology for product category matching in e-commerce. Comput. Ind. **131**, 103497 (2021). DOI URL https://doi.org/10.1016/j.compind.2021.103497

49. Kejriwal, M., Shen, K., Ni, C., Torzec, N.: Transfer-based taxonomy induction over concept labels. Eng. Appl. Artif. Intell. **108**, 104548 (2022). DOI URL https://doi.org/10.1016/j.engappai.2021.104548

50. Kejriwal, M., Szekely, P.: Knowledge graphs for social good: An entity-centric search engine for the human trafficking domain. IEEE Transactions on Big Data (2017)

51. Kejriwal, M., Szekely, P.A.: Information extraction in illicit domains. CoRR **abs/1703.03097** (2017). URL http://arxiv.org/abs/1703.03097

52. Kejriwal, M., Szekely, P.A.: Information extraction in illicit web domains. In: R. Barrett, R. Cummings, E. Agichtein, E. Gabrilovich (eds.) Proceedings of the 26th International Conference on World Wide Web, WWW 2017, Perth, Australia, April 3–7, 2017, pp. 997–1006. ACM (2017). DOI URL https://doi.org/10.1145/3038912.3052642

53. Kejriwal, M., Szekely, P.A.: Neural embeddings for populated geonames locations. In: C. d'Amato, M. Fernández, V.A.M. Tamma, F. Lécué, P. Cudré-Mauroux, J.F. Sequeda, C. Lange, J. Heflin (eds.) The Semantic Web - ISWC 2017 - 16th International Semantic Web Conference, Vienna, Austria, October 21–25, 2017, Proceedings, Part II, *Lecture Notes in Computer Science*, vol. 10588, pp. 139–146. Springer (2017). DOI URL https://doi.org/10.1007/978-3-319-68204-4_14

54. Kejriwal, M., Szekely, P.A.: Constructing domain-specific search engines with no programming. In: S.A. McIlraith, K.Q. Weinberger (eds.) Proceedings of the Thirty-Second AAAI Conference on Artificial Intelligence, (AAAI-18), the 30th Innovative Applications of Artificial Intelligence (IAAI-18), and the 8th AAAI Symposium on Educational Advances in Artificial Intelligence (EAAI-18), New Orleans, Louisiana, USA, February 2–7, 2018, pp. 8204–8205. AAAI Press (2018). URL https://www.aaai.org/ocs/index.php/AAAI/AAAI18/paper/view/16990

55. Kejriwal, M., Szekely, P.A.: Technology-assisted investigative search: A case study from an illicit domain. In: R.L. Mandryk, M. Hancock, M. Perry, A.L. Cox (eds.) Extended Abstracts of the 2018 CHI Conference on Human Factors in Computing Systems, CHI 2018, Montreal, QC, Canada, April 21–26, 2018. ACM (2018). DOI URL https://doi.org/10.1145/3170427.3174364

56. Kejriwal, M., Szekely, P.A., Troncy, R. (eds.): Proceedings of the 10th International Conference on Knowledge Capture, K-CAP 2019, Marina del Rey, CA, USA, November 19–21, 2019. ACM (2019). DOI URL https://doi.org/10.1145/3360901

57. Koren, Y., Rendle, S., Bell, R.: Advances in collaborative filtering. Recommender systems handbook pp. 91–142 (2022)

58. Lewis, T.G.: Network science: Theory and applications. John Wiley & Sons (2011)

59. Li, Y., Choi, D., Chung, J., Kushman, N., Schrittwieser, J., Leblond, R., Eccles, T., Keeling, J., Gimeno, F., Lago, A.D., et al.: Competition-level code generation with AlphaCode. arXiv preprint arXiv:2203.07814 (2022)

60. Lu, Y.: Duke energy used computer vision and robots to cut costs by $74m (2021). URL https://venturebeat.com/2021/07/18/duke-energy-used-computer-vision-and-robots-to-cut-costs-by-74m/

61. Matloff, N., et al.: Immigration and the tech industry: As a labour shortage remedy, for innovation, or for cost savings? Migration Letters **10**(2), 210–227 (2013)

62. Nadeem, M., Bethke, A., Reddy, S.: Stereoset: Measuring stereotypical bias in pretrained language models. arXiv preprint arXiv:2004.09456 (2020)

63. Newman, M.: Networks. Oxford University Press (2018)

64. Peng, D., Zheng, S., Li, Y., Ke, G., He, D., Liu, T.Y.: How could neural networks understand programs? In: International Conference on Machine Learning, pp. 8476–8486. PMLR (2021)
65. Richardson, G.: Wells Fargo: Personalizing real-time conversations with 70 million customers (2022). URL https://www.pega.com/customers/wells-fargo-customer-decision-hub
66. Ruff, K.M., Pappu, R.V.: AlphaFold and implications for intrinsically disordered proteins. Journal of Molecular Biology **433**(20), 167208 (2021)
67. Santos, H., Shen, K., Mulvehill, A.M., Razeghi, Y., McGuinness, D.L., Kejriwal, M.: A theoretically grounded benchmark for evaluating machine commonsense. CoRR **abs/2203.12184** (2022). DOI URL https://doi.org/10.48550/arXiv.2203.12184
68. Shen, K., Kejriwal, M.: A data-driven study of commonsense knowledge using the ConceptNet knowledge base. CoRR **abs/2011.14084** (2020). URL https://arxiv.org/abs/2011.14084
69. Shen, K., Kejriwal, M.: On the generalization abilities of fine-tuned commonsense language representation models. In: M. Bramer, R. Ellis (eds.) Artificial Intelligence XXXVIII - 41st SGAI International Conference on Artificial Intelligence, AI 2021, Cambridge, UK, December 14–16, 2021, Proceedings, *Lecture Notes in Computer Science*, vol. 13101, pp. 3–16. Springer (2021). DOI URL https://doi.org/10.1007/978-3-030-91100-3_1
70. Smith, B.: How TikTok reads your mind (2021). URL https://www.nytimes.com/2021/12/05/business/media/tiktok-algorithm.html
71. Snelgrove, G.: Boeing's flying taxi - what's the latest? (2020). URL https://simpleflying.com/boeing-flying-taxi/
72. Staff, W.: Inside TikTok's algorithm: A WSJ video investigation (2021). URL https://www.wsj.com/articles/tiktok-algorithm-video-investigation-11626877477
73. Szekely, P.A., Kejriwal, M.: Domain-specific insight graphs (DIG). In: P. Champin, F. Gandon, M. Lalmas, P.G. Ipeirotis (eds.) Companion of the Web Conference 2018, WWW 2018, Lyon, France, April 23-27, 2018, pp. 433–434. ACM (2018). DOI URL https://doi.org/10.1145/3184558.3186203
74. Tang, J., Vazirgiannis, M., Dong, Y., Malliaros, F.D., Cochez, M., Kejriwal, M., Rettinger, A.: BigNet 2018 Chairs' welcome & organization. In: P. Champin, F. Gandon, M. Lalmas, P.G. Ipeirotis (eds.) Companion of the Web Conference 2018, WWW 2018, Lyon, France, April 23–27, 2018, pp. 943–944. ACM (2018). DOI URL https://doi.org/10.1145/3184558.3192293
75. Toews, R.: A wave of billion-dollar language AI startups is coming (2022). URL https://www.forbes.com/sites/robtoews/2022/03/27/a-wave-of-billion-dollar-language-ai-startups-is-coming/?sh=1761d1572b14
76. Trinh, T.H., Le, Q.V.: A simple method for commonsense reasoning. arXiv preprint arXiv:1806.02847 (2018)
77. Truby, J.: Decarbonizing bitcoin: Law and policy choices for reducing the energy consumption of blockchain technologies and digital currencies. Energy Research & Social Science **44**, 399–410 (2018)
78. Williams, T., Scheutz, M.: Power: A domain-independent algorithm for probabilistic, open-world entity resolution. In: 2015 IEEE/RSJ International Conference on Intelligent Robots and Systems (IROS), pp. 1230–1235. IEEE (2015)
79. Zhang, T., Subburathinam, A., Shi, G., Huang, L., Lu, D., Pan, X., Li, M., Zhang, B., Wang, Q., Whitehead, S., Ji, H., Zareian, A., Akbari, H., Chen, B., Zhong, R., Shao, S., Allaway, E., Chang, S., McKeown, K.R., Li, D., Huang, X., Sun, K., Peng, X., Gabbard, R., Freedman, M., Kejriwal, M., Nevatia, R., Szekely, P.A., Kumar, T.K.S., Sadeghian, A., Bergami, G., Dutta, S., Rodríguez, M.E., Wang, D.Z.: GAIA - A multi-media multi-lingual knowledge extraction and hypothesis generation system. In: Proceedings of the 2018 Text Analysis Conference, TAC 2018, Gaithersburg, Maryland, USA, November 13–14, 2018. NIST (2018). URL https://tac.nist.gov/publications/2018/participant.papers/TAC2018.GAIA.proceedings.pdf

Augmented Artificial Intelligence

4

4.1 Introduction

Perhaps because the term "Artificial Intelligence" already sounded ominous, and a substitution of human intelligence, an alternate term, *intelligence amplification*, had been proposed as far back as the mid-1950s (e.g., in Ashby's famous work on Cybernetics [4]). This is not a common term anymore, but the premise of its more modern equivalent, *augmented Artificial Intelligence*, is similar.

As the name suggests, augmented AI is meant to improve and extend ("augment") human decision making, rather than outright automate or replace it. According to McKinsey [9], the next decade or two will witness rapid adoption of such technologies, which could radically transform the workplace as people become more symbiotic with machines that are "smarter." This kind of human–machine interaction , which brings out the strengths (or is expected to) of both humans and machines, promises numerous economic benefits if realized, including through improved productivity gains, higher growth of the economy, and better corporate performance. If these benefits are equitably distributed, they would usher in a new era of prosperity and will be sorely needed in societies that are rapidly aging. Indeed, working-age populations in many countries have already been in decline, and despite recent increases in inflation, some prominent economists maintain that long-term trends (and technology itself) are deflationary due to automation [15].

More broadly, augmented AI could also help address "societal" problems, including problems that are preventable due to human error. Recent research has suggested that AI algorithms may be more adept than (or at least within range of), expert human doctors, at diagnosis of diseases from Magnetic Resonance Imaging (MRI) and X-Rays [57]. On the economic front, productivity growth is much needed to foster long-term GDP growth and is ultimately responsible for a country's standard of living being sustained (or growing) over a long period of time. There has been a marked slowdown in productivity growth, a cause of concern for many. Additionally, as the role and utility of AI expand, augmentation is especially

© The Author(s), under exclusive license to Springer Nature Switzerland AG 2023
M. Kejriwal, *Artificial Intelligence for Industries of the Future*, Future of Business and Finance, https://doi.org/10.1007/978-3-031-19039-1_4

expected to help in sectors that have traditionally not benefited much from AI due to a variety of reasons, such as agriculture, media, and pharmaceuticals. For example, in manufacturing, AI could be used for predictive maintenance that is proactive rather than reactive; in e-commerce, it enhances recommendations and optimizes pricing for both consumer and producer in real time; and in fields such as finance and health insurance, identifies fraud more quickly and with fewer false positives or negatives. Ultimately, if augmented AI realizes its promise, the consumer, producer, and worker would all benefit and so would the larger economy.

At the same time, there are many technical challenges around realizing the promises of augmented AI. Some of the challenges are similar to those of "vanilla" AI that seeks complete automation in an ideal lab setting. However, other challenges are unique to augmented AI. For example, visualizing and explaining AI outputs continues to be a hard problem, as is the more social problem of getting people in the workforce to trust the AI and not alienate them. Over the last five years [48], there has been a greater understanding of some of these challenges, but they are far from resolved, and it will likely take at least a decade before corporate culture truly starts shifting toward a mindset that makes full use of technology and humans collaborating together.

4.2   Augmented AI Versus Complete Automation

Augmented AI is often pitted against complete automation (where the human is completely automated away) [44]. In part, this is because of an unfortunate focus on complete automation in the AI research literature, where human studies are rare and usually confined to ground-truth labeling of benchmark datasets. We believe that the dichotomy between augmented AI and complete automation is a false one and that complete automation is, in fact, an *idealization* of the (more pragmatic) augmented AI. To understand this more clearly, imagine two extremes for solving a pattern-matching or predictive problem: completely manual judgment on one end, which is clearly not scalable, and a trained, fully automated system on the other end.

There are few applications where this other extreme truly applies. One reason is that the "test" environment where an AI is actually deployed is usually different from the "training" environment in which it is developed and trained. For example, the test environment is expected to be more "open" where unexpected events and corner cases happen more frequently than were anticipated in the training environment. Even where complete automation was thought to be possible due to the recent successes of deep learning, such as certain natural language processing and computer vision tasks, evidence is starting to emerge that the training set itself might be biased. An infamous example that comes to mind is the work by AlgorithmWatch showing that Google Vision Cloud labeled the picture of a dark-skinned person holding a thermometer as a "gun," while a similar image with a light-skinned person was labeled as holding an "electronic device" [21].

If we require a human being to check each answer that an AI provides, then we are back to the extreme of manual judgment. Making the distinction between augmented and completely automated AI even murkier is the inconvenient fact that there is an entire industry around inexpensive data labeling, usually focused on markets where labor and human rights are not fully enforced, although there are also economic benefits (not dissimilar to those of off-shoring of manufacturing to many developing countries) [26]. Concerning labor rights, for example, an extensive report in Time [40] detailed the working conditions of data labelers in Africa, employed by a company called Sama (which did work with Facebook). According to the report, the workers "perform the brutal task" of viewing (and removing) illegal, or banned, Facebook content before it can be seen by an average user. The Time article covers the outsourcing of content moderation jobs and operations to such countries where labor is cheap, by companies such as Facebook that have extensive data labeling and content moderation needs. Such news generally raises concerns about geopolitical phenomena such as neo-colonialism, with large companies based in Western nations "profiting from exporting trauma along old colonial axes of power . . . toward the developing world."

In defense of Sama, it has sought to pitch itself as an "ethical AI" outsourcing company per the Time report. Also, it claims to have a global workforce, and instances of working conditions in specific countries may not be reflective of the overall picture. Certainly, it claims many important customers on its website, including Google, NASA, and Ford. We do not wade further into the controversy of labor rights in this chapter and refer interested readers to the Time report cited above, but there is no doubt that data labeling is an important industry, and one that is critical to the advancement of AI, especially in the era of deep learning. There is also little dispute that not much attention is paid to it; some researchers have even called it "ghost work". Given that this work is largely done by humans, it begs the question of whether even ordinary AI, as understood in the research community, can be said to support complete automation.

While technical reasons, such as data labeling, for developing augmented, rather than fully automated, AI are important, we must also bear in mind the social and practical realities of the workforce. Workers are hardly likely to be productive if the threat of automation is real and constantly on the horizon. On the other hand, if machines can become valued "partners" of human team members, helping to take rote work and drudgery out of a worker's hands, it frees them up to become more innovative and productive (and arguably, happier) while helping to reduce error rates that inevitably creep up in rote or repetitive tasks.

On the political front, regulators are starting to realize some of the ethical issues that can arise if an AI is left completely to its own devices (e.g., in making financial or hiring decisions). In Chap. 5, we cover relevant regulatory efforts on this front, but it is likely that many such regulations will be proposed in the years to come. Given such political headwinds, AI that is subject to careful human oversight becomes even more attractive [58].

4.3   Key Features and Example Applications

The central feature of augmented AI has already been suggested in the previous section: it assumes some level of human control, especially in cases where the algorithm is likely to go wrong. This opens up some interesting research and implementation issues: which applications (in particular) might benefit from such augmentation and why? And how do we allocate resources, both during training and inference, to human versus machine decision making?

Some of these issues have recently been addressed in the context of a problem with a slightly different name, but the same philosophy: human–machine teaming [31, 34]. According to [34], human–machine teaming may be defined as humans and AI "interacting" with one other as "team members" such that, within the task context, the machine takes on the "teamwork functions" of a (hypothetical) human partner. The unique nature of human–machine teams necessitates a study of fundamental characteristics of such teams that are not completely obvious. Findings from team science may be applicable, but because machines are likely not prone to the same cognitive, social, and emotional issues as human team members (for better or for worse), they need to be thought of differently from a functional standpoint. The synergies are also different: machines have certain strengths, especially if they have been trained in a certain way, and human beings have others. Although much has been written about human–machine teaming in military contexts (with aviation being a good example), much still remains to be discovered in industrial applications. Although we do not distinguish between human–machine teaming and augmented AI for the purposes of this chapter, we do note that the second of the two questions above (resource allocation and optimal utilization of human versus machine capabilities in a human–machine team) is more directly addressed in the human–machine teaming literature than in the augmented AI literature.

To answer the first question of which applications might benefit from augmented AI, a whitepaper published by the Digital Reality group at IEEE [46], an eminent professional society for engineers, provides some guidance and examples. We condense some of the more relevant examples in the list below:

1. Political think tanks, campaign managers, and political action committees (PACs) could use big data analytics and augmented AI to identify undecided voters. However, this kind of analytics can be abused, as the ongoing controversy around Cambridge Analytica demonstrates. Back in the 2016 US presidential election cycle, the political consultancy company was outed by a whistleblower as having used personal data (taken without informed consent or authorization) from as early as 2014, with the goal of building a system that could be used to profile US voters at an individual level. The ultimate intent was to target such voters with "personalized" political advertisements. Indeed, repercussions of this scandal continue to be felt, with Washington D.C.'s attorney general recently suing Facebook founder Mark Zuckerberg personally [50]. Regardless of who should be blamed or held responsible, we cite this example to illustrate a misuse

of AI technology, including augmented AI, where human beings rather than the AI itself is responsible for the misuse.

2. A more responsible use of augmented AI is medical analysis of case files with the intent to more efficiently identify treatment options. Such analysis is useful when the patient has a long or heterogeneous case history, with multiple symptoms, data sources (e.g., electronic health records versus clinical tests), and the potential for an ambiguous diagnosis or treatment. While a medical practitioner should always make the final decision, an AI could help ensure that all of the data is considered holistically and that previous patients' diagnoses and treatments could potentially be used to guide the current patient's needs. In the near future, the use of cheap genetic testing, in tandem with advances in augmented AI and big data analytics, could well lead to the vision of responsible personalized medicine being realized, albeit with the appropriate privacy safeguards in place.

3. With the advent of the Internet of Things (IoT) , smart manufacturing, and robotics, factory automation, overseen by human employees, could be implemented, leading to lower downtime, higher corporate margins, more productive employees, cost efficiencies, and energy optimization. However, as with other uses of augmented AI, it would be incumbent on both governments and corporate entities to ensure that employees that are being replaced on the factory floor have adequate recourse to re-training and up-skilling to qualify them for other jobs. Not doing so would lead to unrest and populism not unlike what we have witnessed in recent years in western nations. IoT could also be used to guide and plan predictive maintenance tasks in factories and industrial workplaces.

4. Advances in natural language processing and chatbots, including "voicebots," could result in more efficient virtual customer assistance, including for increasingly complex needs as more data is gathered through recorded conversations with customers. There has already been a noticeable improvement in speech recognition capabilities of many customer service lines that are initially answered by an automated system. Augmented AI applies because a human being would likely still be needed for the most complex use cases. Even higher-order use cases include techniques such as emotion recognition (through automatic speech analysis), that may be used to direct distressed callers to human beings [3, 24].

The four examples above are only a snapshot of what is truly possible. For example, in finance, robo-advisors are already becoming popular. In the future, we imagine that these advisors will become increasingly human-like, almost like an empathetic financial advisor who understands their client's needs and are able to tailor their advice, and the wording thereof, accordingly. Other applications where augmented AI can play a predominant role, include automation in aviation, and distance education and tutoring. The latter is an especially exciting application area, with the potential to boost educational productivity and reduce spending by tailoring curricula to a student's personalized needs, pace of study, goals, and learning styles. Use of AI in education is actively being explored, including through federally funded efforts by the likes of the US National Science Foundation, although actual deployment of advanced technology in education remains woefully underutilized.

Next, we consider a specific case study in augmented AI (radiology). The remainder of the chapter will be given to an issue that is less technical but much more relevant to long-term strategic planning for companies and their human resource departments: how augmented AI will lead to changes in the workforce.

4.4   A Case Study in Augmented AI: Radiology

Although it is easy to talk about a topic like augmented AI in broad and abstract terms, a case study in a high-revenue domain like radiology helps focus on the issues of interest. Rather than speculate, we relied on peer-reviewed work by actual radiologists in synthesizing the material for this case study, with [30] being the specific primary source for our synthesis.

We picked radiology, as it is one of the medical fields of specialty that has been particularly influenced by AI, both historically and currently. Indeed, as far back as the 1980s, pioneering work had been done in medical imaging perception [25], with obvious applications to radiation safety, as well as other medical imaging and medical physics applications. However, recent advents in deep learning, including convolutional neural networks and the more recently proposed transformer-based neural networks, have led to a dramatic shift in how AI is perceived. Some of the best neural networks, for example, are now more accurate than human radiologists in "narrow" problem areas such as nodule detection [5, 59].

As the authors in [30] point out, radiologists have different strengths than the AI. The latter is good at accurately detecting and classifying images of disease, as well as other features of interest, but it remains poor at explaining its decisions in a rigorous manner, or more generally, in interpreting its detection and classification to make clinical judgments. In contrast, the former has been trained to make just such judgments. Judgment is a complex issue, and not without its own problems of bias, as behavioral psychologists have been quick to point out. Nevertheless, there is a reason that radiologists and other medical doctors spend years in formal medical training and residency. While this suggests we should be confident in their training, it also suggests a limitation: there is a high barrier to entry, and highly trained radiologists, who are able to integrate diverse streams of information from specialists in other areas, free-form discussions, images, and health records, may not be available everywhere in the world. It is not completely clear that even current radiologists are able to handle such an onslaught of "multi-modal" information, especially with new advances and findings in the research literature.

The argument therefore should not be whether humans are better or AI is better. Technology, especially AI technology that has been shown to have high accuracy and validity, should not be resisted irrationally. The authors in [30] argue that the burden is, in fact, on radiologists to further "differentiate" themselves. We hypothesize that good use of augmented AI can help radiologists do just that.

Another reason why use cases such as this are important is that, ultimately, the AI at play here is software, rather than a mechanical or medical piece of equipment (such as a robot, or a better diagnostic device). Traditionally, medical science has often relied on more advanced devices and research to boost outcomes in healthcare, whether it be the invention of better diagnostic equipment,[1] more effective vaccines and therapeutics, or groundbreaking techniques such as CRISPR and mRNA [28, 49], the latter of which played an important role in the rapid development and deployment of COVID-19 vaccines [41].

The observation that AI in this domain is largely software is important because the ability to build and use it lies in the hands of the radiologists (or some consortium of radiologists working together). The neural networks that have super-human accuracy are already publicly available and well-documented. Radiologists who do not have access to the "trained" system need to only train their own, with the help of a reasonably competent computer scientist. In other words, radiologists looking to differentiate themselves and achieve the best outcome for their patients do not need to wait for the next unpredictable (and likely cost-prohibitive) equipment to arrive in their labs: it can be achieved at minimal cost, in-house. By shifting some of the image analysis work to the AI, radiologists can have more face time with their patients and give more cognitive attention to interpretation and clinical judgment. This ultimately results in a win–win for all: radiologists, patients, and hospitals.

In the more realistic medium term, AI can be used to augment the capabilities of radiologists through several task-based use cases. Table 4.1 briefly summarizes these capabilities, as described (more fully) in [30]. While it may seem that an implementation use case such as "intelligence automation" is technically the purest kind of augmented AI in the table, in practice, almost all of the use cases involve augmentation along a spectrum. Decision support systems, for example, never really work in isolation but are like partners to actual radiologists who have to make the final clinical decision (and potentially be legally and medically responsible for such a decision). On the other hand, a task such as "detection and prediction automation" is closer to complete automation.

Many of the implementation use cases noted in Table 4.1 are already being explored for providing practical utility to radiologists. Table 4.2 lists some specific areas constituting a "roadmap" (of sorts) for greater AI augmentation and implementation in radiology. Encouragingly, the areas comprise areas of AI that have historically been disjoint, such as computer vision and data mining. The table shows that, within the scope of a specific domain or complex application, these areas all have a potential role to play. In turn, if the promise of these areas is realized, costs could be driven down for consumers and producers (hospitals, clinics, and medical centers with radiologists) alike, which is going to be critical, especially in an aging society, in a healthcare system that is coming under criticism on all sides for its ballooning costs and inefficiencies.

We end with a note on the ethical issues of implementing AI in a medical specialty like radiology. In general, there are always concerns when implementing

---

[1] As an indicator of how hard even some seemingly "simple" problems, such as doing blood tests on small amounts of blood, can be, one needs to look no further than the infamous saga of the (now dissolved) blood-testing startup, Theranos [56].

Table 4.1 AI implementation use cases (expressed as "task-based" categories) that could be feasibly implemented in radiology [30]

Use case	Description	Further reading
Detection and prediction automation	Among other tasks, computer vision has shown some promise in automating detection of lung nodules on CT scans, and pneumonia on chest X-rays [1,45]. Next steps include predicting the behavior of pre-cancerous lesions on CT scans (e.g., through use of regression and modeling techniques). If this can be done accurately enough, it would reduce the number of unnecessary invasive tests, such as biopsy. There would be immense potential for use in screening the population for cancer, especially in regions of the world where there is a dearth of radiologists on a per-capita basis	[1,45]
Intelligence automation	As a term, *intelligence automation* was first used in the World Economic Forum and seems aligned with what we have presented as augmented AI [12]. In radiology, augmented AI will likely yield higher levels of accuracy in diagnosis, especially in research showing that human–machine teams tend to make more accurate predictions compared to either entity (human or AI) alone. Whether this holds true in general, including in radiology, remains to be seen but is promising and could be evaluated in future peer-reviewed studies. An augmented AI framework would also involve a "radiologist-in-the-loop" to help ensure that patient interests and safety standards are being met. It also ensures legal and ethical compliance and minimizes the probability of moral hazard	[12,37]
Precision diagnostics and Big Data	As research starts to uncover how gene expression is connected to features in tumor images, there will be a need to apply AI to the huge amounts of data generated through imaging to facilitate precision medicine and diagnostics. Precision diagnostics will also be applicable to chronic diseases, including degenerative and neurological disorders, especially in light of an aging society	[11,14, 52,66]
Decision support systems	Although imaging studies have doubled every ten years over the last two decades, the use of AI is not yet ubiquitous, at least in medicine. In non-medical domains, machine learning is already being applied in such critical tasks as driver-assist systems in cars, helping reduce the prevalence of accidents. Driver assist may be thought of as a form of decision support. A similar form of decision support could presumably be applied to diagnostic imaging. This will be especially important during emergencies, or when a study is performed outside of office hours, when only a skeleton-crew is operating. Not operating at such hours would also reduce burnout among doctors and improve radiologists' well-being. Rapid detection of conditions such as stroke, which arise in emergencies, would also be facilitated through neuro-imaging, where AI has already been used to analyze non-enhanced CT and MRI images and shown to automatically detect artifacts such as infarcts, among more advanced tasks (such as differentiating thrombus from plaque in carotid arteries) given CT images as input [19,42]	[19,27, 33,42,55]

Table 4.2 Roadmap for greater AI augmentation and implementation in radiology [30]

Key area	Notes
Image segmentation and lesion detection	The technology is in a ready stage, since it was demonstrated commercially in the 2017 Annual Meeting of the Radiological Society of North America, held in Chicago
Generating reports	This area relies primarily on NLP, rather than computer vision. Its use arises in the fact that many radiology reports are written using long prose rather than as bulleted lists or items. Crafting these reports takes unnecessary hours of typing or dictation and involves significant cognitive burden as the reports must be of sufficient quality (i.e., accurate from both a factual and grammatical standpoint). NLP, especially generative NLP, can help alleviate this load by serving as an augmented AI that creates reports through scanned images, or via speech recognition. A harder, more multi-modal, version of this problem would take both types of data into account (speech and images) and will likely be less accurate
Complex error detection in reports	This is again an NLP-centric task. In essence, the NLP would act as an independent reader or reviewer and would be able to detect errors (both syntactic and semantic) before any report is finalized or disseminated. The technology is fairly mature, e.g., in a study by Mayo clinic, it was found that while 9.7% radiology-based reports generated using speech recognition contained errors, less than 2% could be considered as being consequential or material [47]
Data mining for scientific or clinical research	Many historical radiological reports contain a treasure trove of untapped data. These data are stored in databases of electronic health records, sometimes across the globe. With the proper regulations in place, the data could be "mined" using both knowledge discovery and NLP algorithms to yield valuable insights. One way to do so is to create a "knowledge graph" from this data, making it amenable to structured analysis such as formal querying and predictive analytics. One could imagine, for example, taking such an unstructured database as input, and converting it into searchable databases containing semantically tagged concepts such as disease entities, symptoms, and keywords. If sufficiently mature, the technology could significantly help to speed up, or even automate, entire components of the (currently slow and painstaking) medical research and hypothesis generation pipeline
Business intelligence for radiologists	While many of the previous use cases focused on how machine learning could help make the clinical or knowledge management parts of the job easier, AI can also help to improve business intelligence for radiologists through (for example) intelligent and real-time dashboard, trend analysis, alert systems, workflow management, and performance measurement. In turn, the throughput and efficacy of radiology practices are improved, which would flow through to improved patient satisfaction and other objective measures such as shortened wait times

AI in healthcare. Concerning imaging in particular, the following issues must be noted, derived from the Asilomar AI principles [43]:

- **Safety** is perhaps the most critical imperative, including the safeguarding of the sick individuals' health at a time when they are at their most vulnerable. Although it may seem strange that safety is being cited in the context of imaging, as opposed to administering drugs, vaccines, or surgical medical procedures, the fact is that medical AI systems could influence patient safety through mis-diagnosis or labeling of complex cases, and biased decision making. The black-box nature of neural networks, which are not explainable using usual methods, must especially be borne in mind. The specialist must always remember that not only are AI systems not perfect, but their error distribution is not always random either. Regular checks and validations must be conducted, and one should be cautious about just taking the AI at its word. Medical ethics dictate that doctors should, first and foremost, "do no harm." Therefore, not only must any AI system be put through the most stringent tests, and shown to be safe and accurate before being used on patients, but must be audited on a regular basis.
- **Privacy** is also an important concern in healthcare, since the majority of AI systems would have to have access to health records that are protected, whether the records are on the premises of the healthcare provider or in the cloud. Either way, misuse or mis-handling of such data can pose risks to the privacy of patients. Regulation is important, but relevant legislation needs to be updated to take the role of AI into account. For example, the designers of AI algorithms and systems could ensure that they are only granted access to relevant health records on a case-by-case and need-to-know basis. Furthermore, where third-party AI companies are involved, careful auditing measures must be in place to protect patient privacy and ensure compliance both with the law and medical ethics.
- **Transparency** is important, especially within an AI-centric framework, since it may not always be possible to determine why an AI system failed or otherwise did something "strange." More priority should be placed on such research, and there needs to be less of an overarching focus on accuracy alone. One approach might be to force the model to assign "confidence" scores to each prediction, but a problem with this approach is that neural networks (in particular) have been found to be prone to over-confidence. While statistical methods have been proposed for this problem, the progress is not clear. For example, a recent study at Imperial College, using medical imagery datasets, has shown that advanced confidence detection methods do not necessarily outperform a simple statistical technique [7].
  For AI to truly influence decision making in healthcare, it should be possible to satisfactorily *explain* the processes that it used to arrive at its decision, and such decisions should be amenable to human-in-the-loop processing. Confidence scores alone would not suffice, although they may be part of a broader solution. In this way, not only are human authorities legally liable but it also ensures that radiologists feel that they have agency (and have "veto" power) over what eventually happens.

These ethical concerns aside, influential government agencies such as the Food and Drug Administration (FDA), have signaled that they are relatively eager to support adoption of novel, safe, and effective AI within healthcare [32]. While this is encouraging, we end with the cautionary note that the burden of responsibility for such adoption should always rest with healthcare organizations and with doctors in particular. Public scandals, should they occur, would likely set back progress by decades and (rightfully so) make people skeptical and less trusting that opaque technologies such as AI are being used with their well-being in mind.

4.5 Changes in the Workforce

Unlike many of the other chapters, where we focus on AI as a strategic lever in industries and companies, augmented AI connects directly with the workforce of the future. The connection is in the name itself, since the AI will "augment" the current workforce, but with rising automation, we can also think of the future workforce augmenting the AI itself. This may seem threatening on the surface, since it might suggest that human beings become second-class citizens to the AI in the workforce, but we suggest that it is actually the other way around. A recent report from McKinsey [9], for example, argues persuasively that "higher cognitive" skills will become increasingly in demand, as will social and emotional skills. In contrast, work that is rote is already on the way to being automated, a trend that will likely accelerate in an innovation-driven economy.

The fear that automation will erode jobs and labor's bargaining power is a very real one, and not without justification. Over the last half-century, workplaces have been radically transformed by technology, although fears of technological displacement go back all the way to the industrial revolution. Skills used in many professions have fundamentally changed, and certain jobs have been automated away, while others have been created. Overall, technology has made organizations considerably more productive. Some of these changes can also be studied by comparing official US Department of Labor descriptions of roles. In the past, for example, coal miners were tasked with considerable physical and manual work that relied almost exclusively on motor skills and physical strength. However, there is less reliance on gross physical strength today, when machines do most of the heavy toiling. Instead, more complex skills are brought into play, e.g., the monitoring of equipment and occasional problem-solving.

Yet another example of how workplaces have transformed in nature without the necessary elimination of the labor component itself is the nursing profession. In the mid-1950s, nurses had duties such as monitoring patients by taking their temperature and pulse, administering medicinal doses, and assisting with tasks such as bathing, massaging, feeding, and other therapeutic activities. While some of these tasks, including administering medicines, continue to this day, nurses today are also equipped to do tasks that were commonly only practiced by doctors decades ago, such as performing diagnostic tests and analyzing clinical results.

In a similar vein, bank tellers, who used to perform rote work such as handing cash withdrawals to, or taking deposits from, customers, now find themselves managing customer service issues such as complaints and clarifications, as well as sales issues such as selling and promoting financial products and services. The fact that nurses and bank tellers are able to take on more responsibility attests to increases in labor productivity in these (and other) professional areas.

4.5.1 How Will Organizations Change?

The report by McKinsey suggests that organizations will change in four key ways [9]:

1. First, as companies undergo a "mindset shift" that accompanies any technological change of sufficiently widespread scope (such as the emergence of the gig economy in the last decade), critical to their future success will be in fostering a culture that embraces continuous learning and offers such options throughout the organizational hierarchy.
2. Second, the organization and its human resources (HR) leaders will have to become more comfortable with "diffuse" teams that are much more team-based, collaborative, and cross-functional than today. Specifically, there will be less emphasis on hierarchy and more agile and flexible ways of working. Quite possibly, new business units and HR roles may have to be created, and some roles will likely be eliminated or consolidated.
3. Third, an innovative prediction is that work activities will become more modular, allowing them to be "unbundled" (and consequently, "rebundled"). Similar to how one designs software, with different modules coming together (often requiring an entire team of engineers with different areas of expertise and responsibilities), work activities will be similarly transformed, allowing organizations to make effective and nimble use of different qualifications, expertise, and availability among their workers.
4. Fourth, and perhaps most obviously, the composition of the workforce itself will undergo a radical shift. The trend already began with outsourcing but accelerated more recently with increased dependence on crowdsourcing, freelancing, and, more generally, the growing gig economy. Technologically, companies will start investing more in cloud infrastructure. All of this will have the effect of structurally transforming both traditional labor *and* capital. Companies will become much more asset-light, allowing them to weather business cycles more effectively. However, this change could also be abused, and we are already starting to see some of its impacts in a recent push by workers at the likes of Starbucks and Amazon toward greater unionization, as well as proposed legislation that could accord gig workers greater rights.

The importance of a top-down mindset in embracing such changes, and being proactive rather than reactive, may well determine the long-term prospects of the organization. Augmented AI can be a useful asset in almost all of the above. McKinsey takes particular note of the role that HR will also have to play, especially as competition for top talent intensifies, and skills and job descriptions change with increasing frequency.

The ordering of the organizational changes above was not accidental. Surveys by McKinsey have suggested that continuous learning may indeed be viewed as the most important element that can help both organizations and employees prepare for a changing workforce and labor–capital mix. In Big Tech and innovation-intensive industries, the future is already here. For example, as Google's focus shifted from desktops (as first-class citizens) to mobile, and now AI, the skills of its engineers had to be similarly upgraded. In light of its AI-first culture, the company introduced a training program called "Learn with Google AI". This program offers a paced introduction to machine learning, with the result being that over 18,000 employees (almost a third of its engineering personnel) were able to be trained in this critical area over two years. The course has also been made publicly available, and many thousands of students and practitioners have taken it for free.

Beyond Big Tech, similar innovations are underway, and we suspect the COVID-19 pandemic only accelerated the trend. As one example of how workers in a traditional field will also have to adapt, Rio Tinto (the mining giant) has been adopting autonomous vehicles in some of its mines. Operating such mines will require workers to learn new skills in vehicle repair, maintenance, and operation. Automation, including the use of drone technology, is also being introduced in areas such as agriculture.

Finally, it also bears noting that the physical layout of offices may change to become more "robot"-friendly. Although this sounds like science fiction, some evidence of it can already be found through the likes of companies such as South Korean online platform Naver [62], which recently opened the headquarters of its R&D subsidiary (Naver Labs) in Seongnam in June 2022. Housed in a 28-storey building designated as the "world's first robot-friendly building", the structure is designed to accommodate both humans and machines physically. For example, it will feature a hundred wheeled robots that are primarily designed to deliver packages, as well as food and beverage orders, to the 5000 (human) employees within the building. While the robots would be able to use the 36 human elevators, they will also be able to access a "robot-exclusive" elevator called the Roboport that has a design similar to a Ferris wheel. Robots will be equipped with sensors, such as video cameras, and will also be able to use the building's 5G network. The building also uses other emerging technologies, such as digital twins, which the robots can cross-reference with the camera feed to place them within inches of a target location.

Although Naver seems to be the exception rather than the rule (currently), it is also a harbinger of the future, as such technologies become cheaper and more reliable. It also serves as a good example of how multiple emerging technologies can converge to make a futuristic physical workplace possible.

4.5.2 Demand for Technological Skills

To work with advanced technologies, workers must not only understand how they work but also be able to use them to innovate. With the right experience, such technologies can be developed and adapted to service a diverse set of needs in the workplace. Many occupations today require such skills, well above what was needed even a decade ago, and include data scientists, IT programmers and professionals, designers, and scientific researchers. McKinsey's research suggested that increasing amounts of time will be spent on learning and honing these skills, as companies continue to deploy automation and AI-based technologies such as robotics and advanced analytics. Eventually, as multiple emerging technologies converge (e.g., IoT and AI [53]), the learning curve will become steeper, and novel supporting technologies (such as "no-code" platforms that allow program generation without actual writing of syntactically complex code [13, 38]) will be needed to enhance productivity and find the talent to operate such systems. It is estimated that time spent on learning these advanced technological skills could increase by more than 40% in industrialized nations such as the United States and Western Europe.

The demand mix for specific technological skills is not the same and is likely to fluctuate over time relatively rapidly. Currently, the steepest rise is expected for programming and IT skills; indeed, according to estimates, these could almost double by 2030 compared to a pre-pandemic baseline of 2016. The doubling may occur sooner since the accelerating effects of the pandemic were not anticipated when the estimates were compiled [9]. Augmented AI will continue to become a core, rather than "nice to have," component of many sectors, and in consequence, organizations will have to invest much more in building up and retaining their tech force compared to the past. Top talent will not only be experts in advanced programming that requires robust knowledge of AI paradigms such as deep learning but will also have skills in advanced statistics, data analysis, and mathematics. Others with a more creative bent may be able to marry those skills with technology design and engineering. This relates to the point we made earlier about making work more agile and flexible, including unbundling and rebundling of work specifications to be more productive with fewer specialized workers.

At the same time, it is undeniable that the people with the skills above will only constitute a minority. In the short term, they will be highly in demand and could even stoke inequality, but in the long run, there is reason to believe that basic digital skills and programming will be woven into the fabric of K-12 education. Along with programming and IT, basic digital skills are among the fast-growing category of skills that are needed today. Executive surveys have shown that digital literacy will need to be improved through the company's hierarchy and functional units (including supply chain and middle management, sourcing, procurement, and even HR), and not just among entry-level employees, or those solely involved in programming and IT.

We note that the trends and predictions above are not radically novel; they tend to mark the continuation of existing trends that first emerged in the aftermath of the dot-com bubble. Between 2002 and 2016, research at Brookings identified substantial increase in digital components of occupations that historically were not digital, such as nurses and construction workers [35, 36]. In 2002, for example, only half of occupations had "low" digital requirements, but by the mid-2010s, this number had dropped to 30% and is continuing to slide downward, another trend that will only be accelerated by the effects of the pandemic.

4.5.3 Cognitive Skills and the Future of Work: Is There a Mismatch?

As augmented AI and other emerging technologies continue to make their impact on the workforce, the demand will shift from basic cognitive to higher cognitive skills. Higher cognitive skills include creativity, complex decision making, data-driven information processing, and critical thinking. According to survey-based projections, demand for these skills will grow through 2030 at "double-digit" rates [9]. Contemporaneously, work activities requiring basic cognitive skills will decline. Given the pace at which AI is progressing, basic cognitive skills will all be automated, with AI making many of the decisions that require them, leading to a lot more happening "under the hood" in an organization's technical infrastructure than is currently the case, where humans continue to make many of the decisions.

Of the higher cognitive skills, there is a particular emphasis on creativity in many executive surveys. For some occupations, such as marketing, where consistently developing innovative and "sticky" marketing strategies is what sets apart the top firms and talent, the need for creativity has always been apparent. In recent years, however, there has been a fear and recognition (even in these fields) that too much may have been ceded to technology and special effects, leading to campaigns that do not come across as authentic. While social media influencers, and other creative avenues, can lead to a more grassroots feel for some campaigns, such tactics cannot confer a lasting competitive advantage. AI can help with the technical and visual aspects of creating a good marketing campaign, as evidenced by the rise of "generative" transformer-based deep neural networks that can generate images and write taglines, caption, and even poetry [8]. However, this technology is open to use by anyone in its fundamental form. What is needed, for an organization to set itself apart, is to "train" the model with its own ethos and data so that its outputs cannot be replicated easily outside the firm. Whether such outputs can be protected through other legal mechanisms such as copyrights or patents is a more controversial issue and one that we briefly comment on in the last chapter of the book.

Regardless of how good the technology becomes, the human element of creativity is ultimately what will confer the competitive advantage (at least in the medium term) [39]. Even beyond traditionally creative fields such as copy writing and marketing, creativity will be needed in other functional units of the organization to interpret, keep up with, and predict, market trends, as well as keep pace with a shifting regulatory environment and competitive landscape.

Beyond creativity, higher cognitive skills such as "advanced literacy" and writing, including the ability to read reports and scientific literature by the likes of academics, thought leaders, and think tanks, will also be important, as will be an intuitive and applied grasp of statistical and data analysis skills. In contrast, basic knowledge of statistics, writing, and reading will experience less growth due to widespread availability and use of assistive and augmented AI (i.e., these will not serve as competitive levers anymore, in the same vein as knowledge of word processing or building slide decks is a competitive lever in any high-paying profession at the time of writing). While this does not imply the extinction of, for example, authors or editors, basic aspects of such work will be shifted to machines, while human beings and expert thought leaders will be responsible for generating compelling content.

In practical terms, a lack of growth in basic cognitive skills will likely lead to many back-office functions being automated. Occupations at particular risk include actuarial sciences and accounting, credit scoring, tax calculations, and loan approval and auditing. Computer algorithms and processes implementing robotic process automation or RPA are already starting to take over many of these functions and, while painful for segments of the workforce, are leading to greater efficiencies for consumers and organizations alike. As McKinsey cites in its study [9], in one bank, the process for compiling and reporting quarterly financial results was reduced by more than 60% (from ten days to four), and over 70% of tasks involved were automated. Net reduction of costs was more than 30%. This illustrates a growing shift in labor–capital mix and, in its extreme form, could lead to social unrest. Indeed, in the same vein that automation and technology transformed the blue-collar workplace (for the worse, for those employed in those sectors and jobs) in the early part of the twenty-first century, a similar phenomenon is likely to occur for white-collar jobs over the next two decades. As with so much else, these trends will likely be accelerated due to COVID-19. We note that technology jobs, especially in basic programming and IT, are far from immune from such automation, which has already begun. Already, in May 2022, there is news each week of layoffs and hiring freezes at technology startups and Big Tech alike.

There is also speculation that the encroachment of automation on white-collar jobs will ultimately yield a class of "new-collar" jobs . Whether this will be a net benefit to the workforce is anyone's guess. In the next section, we discuss this phenomenon in greater detail.

4.5.4   New-Collar Versus White-Collar Jobs

In the previous section, we argued that work activities requiring only basic cognitive skills will be particularly hard-hit by AI and automation. Specific skills include basic data processing and input, which were expected (even pre-pandemic) to fall by close to 20% percent in the United States, and by 23% in Europe, in the (roughly) fifteen-year period from 2016 to 2030. Employees in virtually all industrial sectors will likely be affected. Among 25 skills that they studied, McKinsey predicts that these basic data processing skills will be among the largest showing decline. However, "basic literacy, numeracy, and communication" will stay useful (overall) in the medium term. In the long run, however, even those are expected to decline, and advent and improvement of powerful generative AI algorithms such as the famous GPT-3 architecture of OpenAI will likely hasten their demise [8]. Demand for basic literacy as a competitive job function is likely to decline by just over 6% in the United States across the full economy, but by over 25% in sectors such as banking and insurance, where many functions are ripe for automation. In some sectors, demand for these skills will actually increase, especially where a human component is necessary (e.g., in retail and healthcare, demand for communication and basic literacy skills are expected to rise by 12% and 8%, respectively). Once again, we emphasize that these projections were made pre-pandemic. With the digitization and shift to e-commerce and telehealth that have all occurred, and become normalized (post-pandemic), the projections become less certain.

What is the nature of the "new collar" jobs that will be created as white-collar jobs become increasingly automated? As suggested earlier, companies will increasingly start to see work activities as more modular, and amenable to unbundling and rebundling of "atomic" work activities into more complex tasks that changes over time as needs shift. Companies may realize that they could reallocate tasks among workers with different qualification and aptitude levels, e.g., low-skilled activities that were previously executed by their most skilled workers could now get outsourced, or shifted to others whose skills are better suited for those activities at the margin. This type of unbundling and rebundling, which is slowly starting to take effect, would ultimately raise company efficiency, and is hoped to create new middle-skill jobs. A specific example would be that of registered nurses and even physician's assistants, who are now able to do some tasks that physicians used to carry out a few decades earlier (such as administering vaccines or prescribing medicines), even as physicians themselves have more time to talk to patients and handle complex cases. Another example would be that of grid technicians working in utility companies, who are now able to spend more time solving corner-case problems and resolving customer complaints, instead of hand-logging inspection status (or other such administrative duties).

The commonality between these two examples is that, in both cases, tasks are able to be re-allocated from one worker to another (or to a machine) such that skills and time can be made better use of at the margin. Over time, it is quite likely that the efficiencies thereby procured will lead to a lesser need for middle management, many of whose jobs may become automated as a consequence. These examples aside, at least one survey found that 40% of the companies who consider themselves to be extensive adopters of automation expect to shift some tasks currently performed by highly skilled workers to lower-skilled workers. The percentage is significantly less for companies that consider themselves to be moderate or limited adopters of AI but is still well over 10%. In part, this is not due to choice. With productivity already starting to show decline [16], and with rising margin pressure due to inflation and other ongoing economic concerns [18], companies will be under pressure to make the best use of their workforce and to make optimal use of new productivity-enhancing technologies.

Given that rebundling of jobs will require broader, more flexible skills, will education become more, or less, important? Here, the answer clearly depends on the sector. In the high-tech industry, especially in industry labs that are innovating in biotech or AI, there is already an expectation of hiring PhDs. However, there are many emerging new-collar jobs that pay more than the median wage, but that do not require an undergraduate degree, as research has shown [10]. Regardless, unlike previous technologies, AI and automation require both a shift in how capital is deployed (and ROI is calculated) and also a mindset shift in education, training, and employee motivation. Some companies have already started to build up an alternative educational pipeline to train workers with these skills, including partnering with vocational schools (as IBM has done[2]) or foregoing the usual four-year college route for hiring full-time staff, as Google and other tech companies have done [20].

In summary, new-collar jobs that lie at the intersection of white-collar managerial jobs and more traditional blue-collar jobs could well end up redefining the structure of the workforce. Even pure white-collar occupations such as salespeople or mortgage-loan officers will have to get used to working with, and alongside, an AI. Within sales organizations, there is already increasing use of automation for such tasks as lead generation, and identification of opportunistic up-selling. Salespeople who have the "soft" touch, or the X-factor, will be best equipped to work with such leads and opportunities to "close the deal." In contrast, a previous generation of salespeople, who were arbiters or gatekeepers of information, may find that an AI is able to replicate their secret sauce quite effectively. Similarly, mortgage-loan officers will end up spending less time on rote paperwork, and considerably more time handling cases that are exceptions, or otherwise complex.

4.5.5 Adaptation in the C-Suite

As the preceding discussions have made clear, despite significant uncertainties, augmented AI, general automation and other emerging technologies, will have a profound impact on how organizations operate and are structured. For organizations to adapt to these changes, adaptation in the C-suite will be critical. In other words, HR departments and C-suite executives will have to be similarly open to change. Already, there is emphasis on more data-driven decision making. Certainly, an understanding of technology will be a minimum prerequisite for the next generation of leaders and upper management.

There is reason to worry that the current crop of executives, many of whom are still from an older generation where there was less emphasis on such kind of decision making (as opposed to deciding "from the gut"), may not be open to

---

[2] More recently, under CEO Arvind Krishna, IBM has expanded to include more than 170 such academic-industry partnerships with the goal of training over 30 million people worldwide in these new-collar skills by 2030 [23].

such change. In McKinsey's survey [9], almost 20% of the respondents said that their top executives lacked "sufficient understanding of technologies" to lead the organization through the coming changes, including adoption of automation and AI. Such lack of understanding was rated as the second-highest barrier to AI adoption and automation.

It is not clear if this is a self-correcting problem at the level of an individual organization or a team of executives, but we hypothesize that it is self-correcting at the level of a capitalist market. Openness of executives to the potential of adopting and investing in such technology may make the difference to the survival of a company that is not in the strongest position to begin with. This does not mean that the C-suite needs to become a suite of AI experts or that it needs to rely overly on such expertise. It does mean that the executives will need to be aware of the potential of such technologies, and the correct ways in which to measure its ROI (see Chap. 2 for why this can be complex), to facilitate organizational transformation, including the embrace of such models as flexible work functions, emphasis on creativity, and formal incentives for lifelong learning.

We end with a note on the importance of HR in this process . In many ways, HR will play a more important role in fostering, or stymieing, organizational transformations of the nature mentioned earlier. This will manifest both in how employees and candidates are recruited and interviewed, and in how they are trained, evaluated, and promoted. In McKinsey's survey, nearly all business leaders (just shy of 90% percent) indicated their belief that HR functions will have to adapt, at least moderately. There is more controversy on whether HR functions will fundamentally change (36% for "extensive" adopters of automation versus 19% for "limited" adopters). It is telling that, in light of these perceptions, many leading companies are starting to create and promote the role of the chief human resources officer (CHRO) to join the core leadership team. The CHRO is meant to serve a different purpose than an HR director, or head of HR, one that is specifically designed to facilitate the changes required for the future of work. For example, it is expected that the CHRO will delegate the more administrative and logistical aspects of the job to upper management and directors, while making their primary focus the operational and strategic decisions that will be involved in building up and strengthening the future workforce.

4.6 Automation and the Future of Work: Examples from Three Industrial Sectors

In this section, we discuss the role that automation and augmented AI could play in three different, and important, industrial sectors. Although the discussion is brief, by necessity, we generally find that, while the range of required skills varies depending on the sector, the future workforce in all sectors will need to become more flexible and adaptive. This is an important consequence of automation, and one that, despite its social costs, cannot be impeded in the long run. In the 1990s, "emotional intelligence" was touted as being a determining factor (and has since witnessed a resurgence after falling off as a topic of discussion in the early twenty-first century), but today, especially in the wake of the pandemic, flexibility, nimbleness, and adaptability are increasingly being recognized as the element that will differentiate top talent from the rest in an era of increased automation and shifting work responsibilities.

4.6.1 Banking and Insurance

Robo-advisors, fintech, and cryptocurrency have all been hot topics of discussion over the last decade. However, even if we exclude these, financial services have been at the forefront of digital adoption as firms grapple with luring in a new (and disillusioned) generation of millennials and Gen-Z who have become cynical about the usual platitudes of free markets and capitalism. Concerning the future of work, there is little doubt that financial services will see massive skill shifts in the coming years, if not already. For instance, both hedge funds and traditional commission-charging financial advisors are in decline, as automated investing, private equity, and crowd-financed startup investing through platforms such as SeedInvest rapidly take over. Machine learning and AI are at the forefront of innovations such as automated investing and robo-advising. With crunching of ever larger datasets, including satellite and social media data, and open-sourcing of more powerful deep neural network models, it stands to reason that robo-advising will take over all investment decision making over the next decade.

Beyond banking, machine learning is already being enthusiastically applied to insurance, loan underwriting, and fraud detection [17, 51, 54], both because it delivers consistent results (however, it may also introduce bias as a result) and also because it helps organizations trim costs. AI is already being applied in sales and marketing, including personalized advertisements. In coming years, we expect that functions currently the purview of paralegals, sales agents, and insurance underwriters will all be automated, at least in part.

Citing specific statistics, it is expected that automation in banking and finance will lead to just under 40% of the back-office jobs becoming more susceptible to automation, with a potential decrease in total hours worked (by 2030) of 20% or more [9]. Job openings for tellers, brokerage clerks, and accountants, in particular, are expected to significantly decline. Growth is expected in customer interaction occupations, which will require both higher social and emotional intelligence. Needless to say, growth is also expected in the financial industry for AI experts, not just in mathematical "quant" modeling but also AI subfields such as computer vision and natural language processing. One reason is that, as mentioned earlier, finance is increasingly coming to rely on unorthodox sources of intelligence, such as social media and satellite data.

4.6.2   Manufacturing

In an era of "de-globalization" [63] and relocation of manufacturing (both of essentials, but also potentially, of discretionary products) to home markets in the wake of the supply chain crisis of COVID-19 [2], there is considerable potential for AI and automation to add value to the manufacturing value chain. Specific mechanisms include predictive maintenance, real-time production, 3D printing, smart robotics, inexpensive deployment of IoT, and automated supply chains [65].

While employment in manufacturing has obviously been falling in Western nations, in the United States, it started rising again in the mid-2010s, with productivity growing 2.5% or more. This productivity could be sustained, or even increased, with investments in Industry 4.0 through more advanced analytics, efficient allocation of labor and capital, greater realized efficiencies in product development, and increased human–machine collaboration, not just in equipment operation and maintenance, but also marketing and sales, as earlier explained.

Inevitably, this will influence the structure of the workforce, especially in occupations (that are manual) such as assembly workers, currently representing just under half of the employment in the pre-pandemic manufacturing sector. Understandably, the need for physical and manual skills is decreasing overall in the sector at twice the pace as for the economy as a whole. Functional occupations requiring basic cognitive skills, such as routine office support, are also expected to be automated over the next decade.

Growth is expected in the medium term for professional occupations such as engineers, executives, and sales representatives, all of which are aligned with the expected growth in the need for emotional intelligence and social skills. Growth for advanced IT skills and basic digital skills is also expected, albeit the latter may start diminishing over time in line with the findings that were presented earlier. As with other industries, demand for higher cognitive skills is expected to increase, driven by the very real need for more creative and complex information processing skills.

4.6.3   Retail

Emerging digital technologies and augmented AI are also expected to significantly affect labor in the retail sector in the decade leading up to 2030. Already, e-commerce and online channels of marketing, promotions, and sales have become standard, if not dominant, for most major retailers, prompting structural shifts in employment within the industry. For instance, while employees dealing with customer service and interaction issues, including managers and executives, have all experienced rapid growth within retail, office support occupations and those whose skills are routine and manual (e.g., used in activities such as stocking) have been declining, a trend that will accelerate with inexpensive adoption of robotics.

Smart automation technologies and processes, such as self-checkout machines and automatic restocking of shelves by robots, will continue to reshape the economics of retailers, including profits, margins, and inventory [60]. The future is closer than is often anticipated. Already, Amazon has introduced a prototypical "smart store" called Amazon Go that is contactless and cashierless, enabling "walk out" shopping that truly promises to induce structural shifts in the way in which we do physical shopping. The store has been rolled out in several locations in Chicago, New York, and San Francisco and may become more commonplace over the future.

Advances in technology such as self-driving vehicles, drones, and electric trucks may also lead to decline in demand for truck drivers and gig economy workers such as Amazon delivery workers. However, growth will continue to be strong for skills required for inducing customers to find goods and make sales; furthermore, those with IT skills and data scientists are also expected to grow in demand in retail as the sector adopts more data analytics. In general, productivity is expected to significantly increase in the sector, although it is less apparent if the impact on total employment will be positive or negative over the next ten years.

4.7   Conclusion

It is only a matter of time before organizations significantly adopt augmented AI in its various guises, including robotic process automation and AI systems to automate many tasks that middle management and other white-collar professionals are currently doing [61]. Slowly but surely, this is already starting to happen, and in high-tech industries such as Big Tech and biotech, companies are also starting to innovate in how they organize their workforce, and the pipeline through which they recruit their workforce. A still unanswered question, however, is whether AI will lead to greater inequality [29], as certain other capital technologies have unfortunately done, by favoring high-skilled workers over lower-skilled ones, leading to a new technocratic class with fewer numbers, but accounting for a greater proportion of wages and profits. The decline of the blue-collar economy, and decimation of manufacturing jobs, saw the rise of populism, political unrest, and polarization, distrust in "establishment" institutions, and most disturbingly, a rise in authoritarianism [6]. These blue-collar jobs have not been replaced with equally well-paying jobs that make similar use of the same physical and mechanical skills. Research has shown that it is unrealistic, and unfair, to expect coal miners and truckers to suddenly switch in occupation and become coders or software engineers. In our view, the lessons of what has happened over the last 30 years with manufacturing and outsourcing need to be borne in mind rather than ignored (under the superficial, and faulty, assumption that "this time will be different") [64], as the current crop of white-collar jobs also become re-structured into new collar jobs. One must recognize that economic fears are very real, as the sudden uptick in unemployment during the early days of the pandemic made clear. A workforce that is constantly on the lookout for the proverbial pink slip is unlikely to be motivated and take risks, which every innovative process inevitably requires. Nor is it completely evident that the gig economy will come to the rescue, or that technology adoption itself will be seamless and unchallenged. Already, we are starting to see a resurgence of unionization efforts in big companies such as Starbucks and Amazon, and gig economy contractors have started to advocate for greater rights and benefits.

Upper management, consultants, politicians, and HR all need to be mindful of these trends and developments and bear in mind that the company with the long-term view will win in the end. This would imply taking the long, hard (but ultimately, rewarding) road of investing in the workforce by incentivizing lifelong learning [22], re-thinking the nature of work to make it more creative, engaging and suitable for people with different qualifications and skills, and inculcating in current workers, through top-down practice, that AI will augment their work and make it more enjoyable in the end, rather than automate them out of a job.

References

1. Anirudh, R., Thiagarajan, J.J., Bremer, T., Kim, H.: Lung nodule detection using 3D convolutional neural networks trained on weakly labeled data. In: Medical Imaging 2016: Computer-Aided Diagnosis, vol. 9785, pp. 791–796. SPIE (2016)
2. Armani, A.M., Hurt, D.E., Hwang, D., McCarthy, M.C., Scholtz, A.: Low-tech solutions for the covid-19 supply chain crisis. Nature Reviews Materials 5(6), 403–406 (2020)
3. Arora, S.J., Singh, R.P.: Automatic speech recognition: a review. International Journal of Computer Applications 60(9) (2012)
4. Ashby, W.R.: An introduction to cybernetics (1957)
5. Becker, A.S., Marcon, M., Ghafoor, S., Wurnig, M.C., Frauenfelder, T., Boss, A.: Deep learning in mammography: diagnostic accuracy of a multipurpose image analysis software in the detection of breast cancer. Investigative Radiology 52(7), 434–440 (2017)
6. Berberoglu, B.: The global rise of authoritarianism in the 21st Century: Crisis of neoliberal globalization and the nationalist response. Routledge (2020)
7. Bernhardt, M., Ribeiro, F.D.S., Glocker, B.: Failure detection in medical image classification: A reality check and benchmarking testbed (2022). DOI https://doi.org/10.48550/ARXIV.2205.14094. URL https://arxiv.org/abs/2205.14094
8. Brown, T.B., Mann, B., Ryder, N., Subbiah, M., Kaplan, J., Dhariwal, P., Neelakantan, A., Shyam, P., Sastry, G., Askell, A., Agarwal, S., Herbert-Voss, A., Krueger, G., Henighan, T., Child, R., Ramesh, A., Ziegler, D.M., Wu, J., Winter, C., Hesse, C., Chen, M., Sigler, E., Litwin, M., Gray, S., Chess, B., Clark, J., Berner, C., McCandlish, S., Radford, A., Sutskever, I., Amodei, D.: Language models are few-shot learners. CoRR abs/2005.14165 (2020). URL https://arxiv.org/abs/2005.14165
9. Bughin, J., Hazan, E., Lund, S., Dahlström, P., Wiesinger, A., Subramaniam, A.: Skill shift: Automation and the future of the workforce. McKinsey Global Institute 1, 3–84 (2018)
10. Carnevale, A.P., Strohl, J., Ridley, N., Gulish, A.: Three educational pathways to good jobs: High school, middle skills, and bachelor's degree (2018)
11. Collins, F.S., Varmus, H.: A new initiative on precision medicine. New England Journal of Medicine 372(9), 793–795 (2015)
12. Doraiswamy, M., Forget, A.: The real revolution could be IA, 2017
13. Finnie-Ansley, J., Denny, P., Becker, B.A., Luxton-Reilly, A., Prather, J.: The robots are coming: Exploring the implications of OpenAI Codex on introductory programming. In: Australasian Computing Education Conference, pp. 10–19 (2022)
14. Gillies, R.J., Kinahan, P.E., Hricak, H.: Radiomics: images are more than pictures, they are data. Radiology 278(2), 563–577 (2016)
15. Goodhart, C.A.E., Pradhan, M.: The great demographic reversal: Ageing societies, waning inequality, and an inflation revival. Springer (2020)

16. Gordon, R.J., Sayed, H.: The industry anatomy of the transatlantic productivity growth slowdown. Tech. rep., National Bureau of Economic Research (2019)
17. Hall, P., Cox, B., Dickerson, S., Ravi Kannan, A., Kulkarni, R., Schmidt, N.: A united states fair lending perspective on machine learning. Frontiers in Artificial Intelligence **4**, 695301 (2021)
18. Hewson, M.: Inflation and margin concerns put retailers under pressure. (2022). URL https://www.cmcmarkets.com/en/news-and-analysis/inflation-and-margin-concerns-put-retailers-under-pressure
19. Jiang, F., Jiang, Y., Zhi, H., Dong, Y., Li, H., Ma, S., Wang, Y., Dong, Q., Shen, H., Wang, Y.: Artificial intelligence in healthcare: past, present and future. Stroke and Vascular Neurology **2**(4) (2017)
20. Jones-Gorman, J.: Many companies are dropping their jobs' college requirements. (2022). URL   https://www.governing.com/work/many-companies-are-dropping-their-jobs-college-requirements
21. Kayser-Bril, N.: Google apologizes after its vision AI produced racist results (2020). URL https://algorithmwatch.org/en/google-vision-racism/
22. Kejriwal, M.: On preparing for the future of work through proactively inclusive lifelong learning frameworks. Science & Diplomacy (2022). URL https://www.sciencediplomacy.org/perspective/2022/preparing-for-future-work-through-proactively-inclusive-lifelong-learning
23. Kelly, R.: IBM to train 30 million people globally in tech skills by 2030. (2021). URL https://campustechnology.com/articles/2021/10/18/ibm-to-train-30-million-people-globally-in-tech-skills-by-2030.aspx
24. Koolagudi, S.G., Rao, K.S.: Emotion recognition from speech: a review. International Journal of Speech Technology **15**(2), 99–117 (2012)
25. Krupinski, E.A.: The future of image perception in radiology: synergy between humans and computers. Academic Radiology **10**(1), 1–3 (2003)
26. Kshetri, N.: Data labeling for the artificial intelligence industry: Economic impacts in developing countries. IT Professional **23**(2), 96–99 (2021)
27. Lang, K., Huang, H., Lee, D.W., Federico, V., Menzin, J.: National trends in advanced outpatient diagnostic imaging utilization: an analysis of the medical expenditure panel survey, 2000-2009. BMC Medical Imaging **13**(1), 1–10 (2013)
28. Ledford, H.: CRISPR, the disruptor. Nature **522**(7544), 20–25 (2015)
29. Leslie, D., Mazumder, A., Peppin, A., Wolters, M.K., Hagerty, A.: Does "AI" stand for augmenting inequality in the era of covid-19 healthcare? BMJ **372** (2021)
30. Liew, C.: The future of radiology augmented with artificial intelligence: a strategy for success. European Journal of Radiology **102**, 152–156 (2018)
31. Lyons, J.B., Wynne, K.T., Mahoney, S., Roebke, M.A.: Trust and human-machine teaming: A qualitative study. In: Artificial Intelligence for the Internet of Everything, pp. 101–116. Elsevier (2019)
32. Marr, B.: First FDA approval for clinical cloud-based deep learning in healthcare (2017). URL https://www.forbes.com/sites/bernardmarr/2017/01/20/first-fda-approval-for-clinical-cloud-based-deep-learning-in-healthcare/
33. McDonald, R.J., Schwartz, K.M., Eckel, L.J., Diehn, F.E., Hunt, C.H., Bartholmai, B.J., Erickson, B.J., Kallmes, D.F.: The effects of changes in utilization and technological advancements of cross-sectional imaging on radiologist workload. Academic Radiology **22**(9), 1191–1198 (2015)
34. McNeese, N.J., Demir, M., Cooke, N.J., She, M.: Team situation awareness and conflict: A study of human–machine teaming. Journal of Cognitive Engineering and Decision Making **15**(2–3), 83–96 (2021)
35. Muro, M., Liu, S., Whiton, J., Kulkarni, S.: Digitalization and the American workforce (2017)
36. Muro, M., Maxim, R., Whiton, J.: Automation and artificial intelligence: How machines are affecting people and places (2019)
37. Nagar, Y.: Combining human and machine intelligence for making predictions. Ph.D. thesis, Massachusetts Institute of Technology (2013)

38. Paleyes, A., Cabrera, C., Lawrence, N.D.: Towards better data discovery and collection with flow-based programming. arXiv preprint arXiv:2108.04105 (2021)
39. Pasban, M., Nojedeh, S.H.: A review of the role of human capital in the organization. Procedia-Social and Behavioral Sciences **230**, 249–253 (2016)
40. Perrigo, B.: Inside Facebook's African Sweatshop (2022). URL https://time.com/6147458/facebook-africa-content-moderation-employee-treatment/
41. Polack, F.P., Thomas, S.J., Kitchin, N., Absalon, J., Gurtman, A., Lockhart, S., Perez, J.L., Marc, G.P., Moreira, E.D., Zerbini, C., et al.: Safety and efficacy of the BNT162b2 mRNA Covid-19 vaccine. New England Journal of Medicine (2020)
42. Prevedello, L.M., Erdal, B.S., Ryu, J.L., Little, K.J., Demirer, M., Qian, S., White, R.D.: Automated critical test findings identification and online notification system using artificial intelligence in imaging. Radiology **285**(3), 923–931 (2017)
43. Principles, A.: Future of Life Institute. Retrieved January **14**, 2019 (2017)
44. Raisch, S., Krakowski, S.: Artificial intelligence and management: The automation–augmentation paradox. Academy of Management Review **46**(1), 192–210 (2021)
45. Rajpurkar, P., Irvin, J., Zhu, K., Yang, B., Mehta, H., Duan, T., Ding, D., Bagul, A., Langlotz, C., Shpanskaya, K., et al.: CheXNet: Radiologist-level pneumonia detection on chest X-rays with deep learning. arXiv preprint arXiv:1711.05225 (2017)
46. Reality, I.D.: What is augmented intelligence? (2019). URL https://digitalreality.ieee.org/publications/what-is-augmented-intelligence
47. Ringler, M.D., Goss, B.C., Bartholmai, B.J.: Syntactic and semantic errors in radiology reports associated with speech recognition software. Health Informatics Journal **23**(1), 3–13 (2017)
48. Roski, J., Maier, E.J., Vigilante, K., Kane, E.A., Matheny, M.E.: Enhancing trust in AI through industry self-governance. Journal of the American Medical Informatics Association **28**(7), 1582–1590 (2021)
49. Ross, J.: mRNA stability in mammalian cells. Microbiological Reviews **59**(3), 423–450 (1995)
50. Rushe, D., Milmo, D.: Zuckerberg sued by DC attorney general over Cambridge Analytica Data Scandal (2022). URL https://www.theguardian.com/technology/2022/may/23/mark-zuckerberg-sued-dc-attorney-general-cambridge-analytica-data-scandal
51. Sachan, S., Yang, J.B., Xu, D.L., Benavides, D.E., Li, Y.: An explainable AI decision-support-system to automate loan underwriting. Expert Systems with Applications **144**, 113100 (2020)
52. Samala, R.K., Chan, H.P., Hadjiiski, L., Helvie, M.A., Kim, R.: Identifying key radiogenomic associations between DCE-MRI and micro-RNA expressions for breast cancer. In: Medical Imaging 2017: Computer-Aided Diagnosis, vol. 10134, pp. 200–206. SPIE (2017)
53. Shah, R., Chircu, A.: IoT and AI in healthcare: A systematic literature review. Issues in Information Systems **19**(3) (2018)
54. Shrivastava, A.: Usage of machine learning in business industries and its significant impact. Int. J. Sci. Res. Sci. Technol **4**(8) (2018)
55. Smith-Bindman, R., Miglioretti, D.L., Larson, E.B.: Rising use of diagnostic medical imaging in a large integrated health system. Health Affairs **27**(6), 1491–1502 (2008)
56. Straker, K., Peel, S., Nusem, E., Wrigley, C.: Designing a dangerous unicorn: Lessons from the Theranos case. Business Horizons **64**(4), 525–536 (2021)
57. Ursuleanu, T.F., Luca, A.R., Gheorghe, L., Grigorovici, R., Iancu, S., Hlusneac, M., Preda, C., Grigorovici, A.: The use of artificial intelligence on segmental volumes, constructed from MRI and CT images, in the diagnosis and staging of cervical cancers and thyroid cancers: a study protocol for a randomized controlled trial. Journal of Biomedical Science and Engineering **14**(6), 300–304 (2021)
58. Verdiesen, I., Santoni de Sio, F., Dignum, V.: Accountability and control over autonomous weapon systems: A framework for comprehensive human oversight. Minds and Machines **31**(1), 137–163 (2021)
59. Wang, H., Zhao, T., Li, L.C., Pan, H., Liu, W., Gao, H., Han, F., Wang, Y., Qi, Y., Liang, Z.: A hybrid CNN feature model for pulmonary nodule malignancy risk differentiation. Journal of X-ray Science and Technology **26**(2), 171–187 (2018)

60. Wankhede, K., Wukkadada, B., Nadar, V.: Just walk-out technology and its challenges: A case of Amazon Go. In: 2018 International Conference on Inventive Research in Computing Applications (ICIRCA), pp. 254–257. IEEE (2018)
61. Wesche, J.S., Sonderegger, A.: When computers take the lead: The automation of leadership. Computers in Human Behavior **101**, 197–209 (2019)
62. Wilson, M.: Futuristic office was designed for 5,000 people and 100 robot coworkers (2022). URL https://www.fastcompany.com/90754724/this-futuristic-office-was-designed-for-5000-people-and-100-robot-coworkers
63. Witt, M.A.: De-globalization: Theories, predictions, and opportunities for international business research. Journal of International Business Studies **50**(7), 1053–1077 (2019)
64. Yeo, C., Saboori-Deilami, V.: Strategic challenges of outsourcing innovation in global market. Asia Pacific Journal of Innovation and Entrepreneurship (2017)
65. Younis, H., Sundarakani, B., Alsharairi, M.: Applications of artificial intelligence and machine learning within supply chains: systematic review and future research directions. Journal of Modelling in Management (2021)
66. Zhu, Z., Albadawy, E., Saha, A., Zhang, J., Harowicz, M.R., Mazurowski, M.A.: Deep learning for identifying radiogenomic associations in breast cancer. Computers in Biology and Medicine **109**, 85–90 (2019)

AI Ethics and Policy

5

5.1   Introduction

As any new technology starts to develop to the point where it is being integrated broadly across society, formulating and addressing the ethical and legal challenges of the technology rapidly becomes a priority. AI is no different. Prior to the mid-2010s, before deep neural networks really started to make their impact felt across multiple applications and industries, discussions on AI ethics and policy were largely relegated to niche conferences and academia fora. Even within computing, discussion on AI ethics was not prominent. To take just one simple example, for a long time, ethics statements were neither expected nor required in papers published in the major AI or machine learning conferences (unlike papers in healthcare, medicine, or epidemiology).

The current era could not be more different and, as we later argue, can be traced to both governmental and academic focus on AI in major economies in the mid-2010s. Today, the topics of AI ethics, policy, and regulation are intertwined and have prominent conferences and journals dedicated to them in academia (along with academic centers studying the subject in highly ranked universities), as well as policy groups and white-papers instituted by regional governmental bodies such as the European Commission.

In fact, as we discuss in depth later in the chapter, there is now concerted effort by governments across the world to legislate on (or at the very minimum, bring to the floor) issues of AI ethics. In an article published in February 2022, Fortune reported on how, 'buried' in the National Defense Authorization Act for the 2022 fiscal year signed by President Joe Biden, were the Artificial Intelligence Capabilities and Transparency Act (AICT) and the Artificial Intelligence for the Military Act (AIM). These two acts suggest that, at least at the federal level, the US government not only recognizes the importance of AI as a strategic and transformative technology (with the potential for misuse) but is requiring AI ethics, and a continued understanding

© The Author(s), under exclusive license to Springer Nature Switzerland AG 2023
M. Kejriwal, *Artificial Intelligence for Industries of the Future*, Future of Business and Finance, https://doi.org/10.1007/978-3-031-19039-1_5

of AI, as a fundamental and integral requirement of the nation's national security, including within the military, and the Department of Defense (DoD).

A telling insight yielded by the AICT is the definition of AI ethics itself as "the quantitative analysis of artificial intelligence systems to address matters relating to the effects of such systems on individuals and society, such as matters of fairness or the potential for discrimination." Although this definition captures many of the essentials, it cannot be held to be an officially agreed-upon definition, because no such definition currently exists (at least, to the best of our knowledge). The field of AI is still a young field from a scientific point of view, and AI ethics and policy are newer still.

This chapter is not primarily about doing a "deep dive" on AI ethics and policy. Rather, we seek to illuminate some broad trends that should be on the radar of any business, especially one that is looking to make a new foray into AI and that may be international in scope. Simply put, such trends and practices cannot be ignored any longer for any business or industry looking to have even minor involvement with AI. Promisingly, AI regulations were once in the wild west but are starting to have shape and consistency, especially in the major nation-states and Western entities such as the European Union. An analog can be found in the financial industry: while regulations are complex, there is a large degree of international co-operation, allowing many financial institutions to have an international footprint. To provide specific contexts and examples, we also supplement the broad trends with two case studies, one of which is more recent and US-centric (such as the AICT) and the other of which is Euro-centric but has been internationally influential in the space of privacy rights, which is intimately connected to some of the ethical concerns raised about (unregulated) AI.

5.2   AI Versus Digital Ethics

Although this chapter is primarily concerned with AI ethics, it is important to both distinguish, and understand the commonalities, between *digital ethics* and *AI ethics*. In a pragmatic discussion of digital ethics, we would define "digital" technologies as those that encompass *emerging* technologies (i.e., since the early 2000s) that apply, or have substantive dependence on, developments and advancements in computer science and engineering. Such emerging technologies include not only AI and machine learning (ML), which are the foci of this book and chapter, but also blockchain, Big Data analytics, cloud computing, and Internet of Things (IoT).

These areas are not mutually exclusive, and in some cases (especially Big Data analytics), they may even have significant overlap with AI and ML. One may even think of Big Data analytics as being a product of AI and Big Data. On occasion, and in some industrial literature, these digital technologies are conceived of as key contributors to a "fourth Industrial Revolution" that aims to fuse technologies that blur the boundaries between digital, physical, and biological technologies, and that include such applications as virtual and augmented reality, neural-machine interfaces, robotics, and cyberspace. Clearly, any discussion of ethics that applies *broadly* to these technologies would also apply *specifically* to AI, and in some (albeit, not all) cases, the converse would also hold. Where possible, we aim to emphasize such generalities in this chapter, but our primary focus will be on ethics and privacy as they pertain to AI and ML technologies, applications, and advancements.

Similar to conceptualizations of digital ethics that consider social, environmental, political, and psychological perspectives, we seek holistic perspectives in our treatment of AI ethics. Here, a psychological perspective of AI ethics considers *agency* (i.e., issues of moral control and self-determination), including cognitive aspects such as social media addiction, reasonable protection from manipulation and deception, and social contagion. A social theme would refer to themes of identifying with, and belonging to, communities. Environmental themes include not only the carbon footprint of mass device manufacturing but also the enormous amount of energy that it often takes to train some of the latest deep learning models. Such themes also apply to non-AI digital technologies such as blockchain and bitcoin mining. A political perspective is typically related to legal, democratic, jurisdictional, and even economic discussions of AI, including geopolitical tensions and fears (or perceptions thereof) that may arise between established and emerging loci of power, such as the United States and China, due to divergence in ethics, labor force expertise and training, data availability, government incentives, and legal frameworks.

Often, these perspectives can intersect. For example, if a law is passed that gives every individual the right to know when they are interacting with a "non-human" entity (such as a bot), then the ethical perspective needs to take both the psychological and political (and perhaps even the social) themes into account. Because of the complexity of these problems, and the high level of domain expertise required, AI ethics is a rapidly evolving and interdisciplinary area of study. We do not believe in making a strong (or artificial) distinction between AI and digital ethics; many of the issues that apply to one will also apply to the other, with the understanding that "digital" encompasses other emerging technologies that are not necessarily AI.

Finally, we note also that any discussion of ethics on specific issues (regardless of perspective) must also consider the scope of the impact. Political impact sometimes receives the most attention in the press and in elected bodies such as the US Congress, but social and psychological impacts can have much more tangible and long-term impacts on everyday people. At the same time, it is undeniable that echo chambers and misinformation, which can both be fueled by AI, can have lasting effects on the socio-political cohesion of the country and lead to a level of political partisanship and division that has rarely been witnessed in recent decades.

5.3   The Philosophy of Ethics: A Brief Review

Before delving into AI ethics, we briefly review the *philosophy* of ethics. Ethics is a vast field of study, and the scope of this study is enormous, with many schools and traditions that have evolved over millennia across diverse cultures. There are three major schools of thought that are relevant to the discussion at hand, however. The first of these is called *utilitarianism* or *consequentialism* and is often thought of as the ethics underpinning distinctly Western traditions such as common law. An ethical framework that relies on utilitarianism involves the construction of rules and principles that aim to maximize the expected *pleasure* ('utility'), or equivalently, minimize expected displeasure.

Interpretations of pleasure, and the way in which the expectation ought to be calculated (even in principle) and quantified, lead to different interpretations and applications of utilitarianism. From an operative standpoint, this approach to ethics justifies government policies, incentives, and decision making by relying on the view that such policies would maximize some "measure of pleasure" (such as subjective well-being, Gross National Product, and so on) or minimize other "measures of displeasure" (such as income inequality, suicide rate, poverty rate) and should thereby be implemented and enacted. In many cases, there is a clear tradeoff between maximizing or minimizing some of these measures, which leads to policy and political disputes. For example, maximizing economic growth and GDP can sometimes be at odds with minimizing the income inequality in a national context. The differing philosophies of many political parties, especially in Anglo-American nations like the United States, can often be framed in terms of such tradeoffs, although the arguments made in political speeches and campaigns tend to be less direct.

The second relevant school of thought in ethics (for the purposes of this chapter) is *rights-based ethics* that underpins much of European continental law. According to this school of thought, a person is entitled or is conferred "rights" because they belong to a class. Although this may initially seem to have a negative connotation and did have a practical negative connotation (at least historically, for disenfranchised groups, including women, minorities, and subjects of colonialism), much depends on what we mean by "class." When reference is made to "human rights," for instance, the class is understood to be humanity itself. Another example that is oft cited is "civil rights," which are accorded to individuals who are members of the politic.

The third, and final, relevant school of thought in mainstream ethics is *virtue ethics*, which is also known as *natural ethics*. In Western civilization, virtue ethics has its roots in Aristotle, while in the East, it is traceable to Confucius and Mencius, among others. This view of ethics focuses on character development, with a person tending toward an idealized version of themselves over time. Good character is associated with virtues that include, but are not limited to, courage, fairness, integrity, and honesty. In principle, virtue ethics is fairly intuitive. However, a full treatment, including "pros and cons," requires a much deeper knowledge and reading of philosophical sources. For instance, similar to the other schools of thought, virtue ethics can itself be sub-divided into eudaimonist virtue ethics, agent-based virtue ethics, Platonistic virtue ethics, and target-centered virtue ethics. For a survey of these, as well as objections to virtue ethics, we refer the interested reader to an accessible, yet authoritative, encyclopedic source [1].

As the Stanford Encyclopedia of Philosophy, and other sources, duly note, any serious and plausible ethical theory that is normative by design and that considers a range of real-world situations (or in our case, conception, design and use of AI technologies) must also consider elements from all three theories. Put simply, a virtue ethicist would not just attend to "virtues," any more than utilitarians would attend to consequences. However, there are also important obvious differences between the three schools, and sometimes, one school will provide a better framework than the others through which to view an issue. One example of such a difference is that, while utilitarians will *define* virtues as characteristics that yield good consequences, a virtue ethicist would likely resist the urge to define virtues using other "fundamental" concepts; indeed, virtues themselves are considered to be the fundamental units for grounding the other concepts and theories.

Insofar as AI and almost any digital technology goes, policy-making in the short term often tends to consider utilitarian grounds, but rights-based ethics can play an important role when considering legislation and regulation from a constitutional lens. In the United States, for example, there is an ongoing debate over whether social media platforms constitute a "public forum" and whether curtailing any kind of information dissemination (even if it is misinformation) constitutes a violation of first amendment (i.e., free speech) principles in the US constitution. It is all but inevitable that some of these issues, currently being litigated in the lower courts, will eventually have to be decided by arbiters such as the US Supreme Court, or through appropriate legislation (which may itself be subject to litigation). Utilitarian principles and rights-based ethics may come into conflict in deciding such cases, and a satisfactory resolution remains to be seen.

5.4   AI Ethics in Policy

As AI and its applications continue to proliferate in society and industry, its intersection with ethics becomes ever more prominent. AI ethics is still very much in its infancy, despite much impressive recent scholarship that we cite in this chapter. In part, the need for AI ethics, and its policy-centered pragmatic counterpart of AI policy, arose due to a series of high-profile cases where the AI caused harm, sometimes because it had intrinsic design flaws, and in other cases, because of its misuse or even abuse. The former includes biases in AI models that have real-life consequences such as rejections of loans or medical diagnoses. The latter includes a wide range of cases, including voter manipulation, "deepfake" generation, and surveillance. Below, we briefly describe some potential instances, stemming both in industry and law enforcement (and we note, not necessarily by *intent*), illustrating thorny issues that can arise with implementing AI without being mindful of full ethical ramifications.

1. In 2016, a report by ProPublica described how a system called COMPAS, which was used to predict recidivism in Broward County in Florida, tended to incorrectly label African-American defendants as "high-risk" at almost twice the rate that white defendants were thus mislabeled [3].
2. Research has also found that stereotypes can get reinforced by some algorithms. For instance, facial recognition technologies have been found to exhibit differential error rates by race and gender [4]. Researchers found (in one example) that when executing a search using a phrase such as "CEO image search," only 11% of the top results for "CEO" were women, even though in the actual population at the time, 27% of US CEOs were women [10].
3. Recently, the US federal government issued a warning that AI screening software could be used to discriminate against job candidates [9]. The warning was meant to put employers on notice that if said discrimination was occurring, it could put them in violation of civil rights laws. Such software includes AI resume scanners, game-like tests that are administered online to assess job skills, and video interviewing tools that measure an individual's facial expressions and speech patterns. Although such technology seems useful at first glance, it could be used to discriminate against individuals with speech impediments and other disabilities such as arthritis (or other physical and mental impairments). As the warning was issued recently this year, it remains to be seen whether further evidence or lawsuits will be forthcoming.

These cases suggest that some form of oversight is necessary to rein in AI technologies and their infringement on individuals' rights and liberties. Alarmingly, there may be cases that were not picked up by the press, or quietly "resolved" without a true reckoning of any harms they may have caused. It is certainly not the case that the cases collectively covered by the press or in academic literature form an exhaustive set. Structurally and from a policy standpoint, the uses and abuses of AI can be thought of at three levels: national, regional, and international. By regional, we do not mean municipal or local; rather, we refer to regulation that is at the level of several countries with common goals. Although the ethical ramifications of AI (as well as the speculative ethical ramifications of a hypothetical "AI singularity") are still being worked out, policy-makers at these three levels are becoming increasingly alert to its uses and abuses. We provide a few examples below:

- **United Kingdom (UK):** Within the UK, AI application and development has been supported via initiatives such as the Digital Economy Strategy in the mid-2010s. More recently, in 2021, a 10-year National AI Strategy was also unveiled, describing actions for assessing AI risks that are long-term, including catastrophic risks that could be posed by (for example) Artificial General Intelligence.

- **China:** Regulation of AI in China is probably best understood through a document titled "A Next Generation Artificial Intelligence Development Plan" issued by the State Council in 2017. In that document, the State Council and the Central Committee of the Communist Party of China exhort the governing bodies of China to promote AI development. Although regulation of ethical and legal dimensions of AI development is still in its infancy, the policy does ensure state control of Chinese enterprises and over data deemed valuable or sensitive, including storage of Chinese user data within the country. Also, use of, and compliance with the national standards for AI issued by the government is mandatory. These standards do not just encompass AI as understood, but also over big data, industrial software, and cloud computing. In 2021, a set of published ethical guidelines by the Chinese government stated that researchers should ensure that AI does not endanger public safety, is within human control, and complies with human values.

- **United States:** Similar to the other countries discussed above, the United States began a concerted effort to better understand the risks and regulations for AI in the mid-2010s. In 2016, the National Science and Technology Council under the Obama administration released a report titled "Preparing For the Future of Artificial Intelligence" that sought to set a precedent that would permit researchers to continue developing new AI technologies with fewer restrictions. The report states that "the approach to regulation of AI-enabled products to protect public safety should be informed by assessment of the aspects of risk" and that these risks would form the primary basis on which to create new regulation, assuming that existing regulations fail to apply to AI technology.

  The next major discussion on this issue occurred in early 2019, when the Office of Science and Technology Policy (OSTP) within the White House released a draft *Guidance for Regulation of Artificial Intelligence Applications*. This followed on the heels of an Executive Order from the White House on *Maintaining American Leadership in Artificial Intelligence*. The draft guidance contains 10 principles for US federal agencies that could help them decide whether to regulate AI (and if so, how). Following this guidance, the National Institute of Standards and Technology (NIST) released a position paper, but they were not alone: the National Security Commission on Artificial Intelligence (NSCAI) followed by publishing a report, and the Defense Innovation Board also issued recommendations on ethical usage of AI. In 2020, the Trump administration also called for comments on regulation in another draft of the previously issued guidance. We cite these specific instances to illustrate how, in this short period of time, a burst of activity (not unlike what was happening in other nation-states) has followed on AI regulation. Another example of an agency that has been working on AI regulation include the Food and Drug Administration (FDA), which is aiming to create pathways that allow responsible incorporation of AI in medical imaging technology and systems.

- **Canada:** The Pan-Canadian Artificial Intelligence Strategy was created in 2017 and has federal support to the tune of $125 million Canadian dollars. The strategy has the stated objectives of (i) increasing the numbers of leading-edge

AI researchers and skilled graduates in Canada; (ii) establishing "nodes" of scientific excellence at the three major AI centers in Canada; (iii) developing "global thought leadership" on multiple aspects of AI advances, including ethical, economic, legal, and policy implications; and (iv) supporting a national research community working on AI.

A keystone of the strategy is the Canada CIFAR AI Chairs Program that benefits from funding to the tune of (Canadian) $86.5M over a half-decade period to attract and retain the best AI researchers in the world. In 2019, the Canadian government appointed an Advisory Council on AI to explore how AI advancements could best reflect Canadian values, which include human rights, openness, and transparency. In 2020, the federal government and Government of Quebec jointly announced the opening of the International Centre of Expertise in Montreal for the Advancement of Artificial Intelligence. This Centre will advance the responsible development of AI. The Centre also has an international basis in that it has a relationship with the Global Partnership on AI, as we subsequently discuss.

- **Regional efforts in the European Union (EU):** Moving from individual nation-states to regional efforts instituted for regulating AI and forming policy, the European Union (EU) is probably the best known example and is guided by a *European Strategy on Artificial Intelligence*. This strategy is supported by an expert group on AI. In April 2019, the European Commission published its *Ethics Guidelines for Trustworthy Artificial Intelligence*, followed by issuance (in June 2019) of recommendations for policy and investment in trustworthy AI. The Commission's expert group, called the *High Level Expert Group on Artificial Intelligence* , carries out this policy-level work on trustworthy AI, with the Commission itself issuing reports on such matters as the *Safety and Liability Aspects of AI* and on the *Ethics of Automated Vehicles*. As with many nation-state initiatives, discussions on these matters are ongoing but have arguably become more prominent[1] in the late 2010s.

Just before the COVID-19 pandemic would become a major international concern, in February 2020, the Commission also published a white paper titled *Artificial Intelligence—A European approach to excellence and trust*. This white paper comprises two building blocks based on building an "ecosystem of excellence" and an "ecosystem of trust". While the latter outlines the EU's approach for a regulatory AI framework, including explicitly differentiating between "high-risk" and "non-high-risk" AI applications, the former would properly be in scope for a future EU regulatory framework.

High-risk applications would be guided by criteria such as whether the application is being employed in *critical sectors* or otherwise has *critical use*. Developing or deploying high-risk AI applications would include imposing specific requirements for training data; proper standards for data and record-keeping; the so-called informational duties; rigorous requirements for ensuring

---

[1] For example, in 2020, the Commission solicited views on a proposal for AI-specific legislation.

robustness and accuracy; inclusion of human oversight; and application-specific requirements, including for AI systems that are used for remote biometrics. Non-high-risk applications (which may not necessarily be 'low-risk') would have more leeway from a regulatory standpoint. Such applications could, for instance, be governed by a voluntary labeling scheme.

In January 2021, the Commission published their official "Proposal for a Regulation laying down harmonized rules on artificial intelligence (Artificial Intelligence Act)" although a draft had already been leaked a week earlier, and in short order, the Artificial Intelligence Act was formally proposed. One of the concerns that has been noted, especially by private organizations and industry, is the extensive proliferation of proposed legislation on a rapidly evolving field by the Commission. It is undeniable that, at least in part, the speed of these initiatives is guided by the EU's political ambitions in this space, and if not implemented or conceived properly, could have unintended consequences (both for citizens and for innovation).

The Commission itself has stated that it is guided by objectives of strategic autonomy and digital sovereignty. However, the extent and scope of these objectives, or even their definitions, are not completely clear, especially to outside observers. We note that such matters are far from settled, but the Commission's deliberations in this space provide a good example of regional initiatives to regulate AI.

- **Council of Europe (CoE):** Regionally, the Council of Europe (CoE) is an international organization comprising 47 member states, which aims to promote human rights, democracy, and the rule of law. The CoE includes all 29 signatories of the EU's 2018 *Declaration of Cooperation on Artificial Intelligence*. The CoE has created a shared legal practice wherein members have legal obligations to guarantee rights as set out in the European Convention on Human Rights. AI is explicitly mentioned: "The Council of Europe's aim is to identify intersecting areas between AI and our standards on human rights, democracy and rule of law, and to develop relevant standard setting or capacity-building solutions." To this effect, germane documents that have been identified by the CoE include charters, strategies, guidelines, papers and reports and strategies, with authoring bodies not confined to any one sector of society. Such bodies are multi-stakeholder and include organizations, nation-states, and private companies.

- **Global Partnership on Artificial Intelligence (GPAI):** Launched in June 2020 with 15 members, the stated need for the Global Partnership on Artificial Intelligence (GPAI) is for AI to be developed in a manner that aligns with democratic values and human rights. Such alignment is necessary to foster public confidence and trust in the technology. This is also outlined in the Organization for Economic Co-operation and Development (OECD) Principles on Artificial Intelligence [12]. Founding members of the GPAI include Australia, Canada, the European Union, France, Germany, India, Italy, Japan, Rep. Korea, Mexico, New Zealand, Singapore, Slovenia, the USA, and the UK, with the Secretariat hosted by the OECD in Paris, France.

The GPAI mandate covers four themes. Two of these themes (*responsible AI* and *data governance*) are supported by the International Centre of Expertise in Montreal for the Advancement of Artificial Intelligence.[2] In the wake of COVID-19, the GPAI also explored the possibilities for beneficially utilizing AI to respond effectively to the pandemic.

- **Other global initiatives:** The GPAI is a good example of a global initiative that takes an international and multi-stakeholder approach to aligning AI with responsible and beneficial use, but it is not the only one. We already mentioned the OECD Recommendations on AI that were adopted in May 2019. Another example includes the G20 AI Principles adopted in June 2019. Later in 2019, the World Economic Forum also issued 10 "AI Government Procurement Guidelines." In February 2020, the EU also published a draft strategy paper for regulating AI, while also promoting it at the same time.

  Within the United Nations, there has been much discussion on AI regulation and policy, including at the UNICRI[3] Centre for AI and Robotics. In its 40th scientific session in late 2019, UNESCO[4] began a two-year process that sought to achieve a "global standard-setting instrument on ethics of artificial intelligence." With this goal in mind, a number of UNESCO conferences and fora have been held specifically to elicit the views of stakeholders in this matter. A draft of the recommendations on AI ethics by the UNESCO Ad Hoc Expert Group was released in September 2020 and included calls for legislative gaps to be addressed.

We conclude this discussion with several common observations. First, as many of the instances suggest, major nation-states were already starting to pay attention to AI ethics and regulation in the mid-2010s. No doubt the media had an important role to play in bringing this about, since several documented successes of AI were being touted in the press at the time. In large part, this was because deep learning had demonstrated its utility in academic circles and Big Tech, and there was a race to adopt it in other firms and industrial sectors as well. Startup activity was also becoming more pronounced.

By 2018–2019, with the advent of transformer-based models such as BERT, and other deep learning systems (including deepfake technology), concerns about AI could not be ignored. This may explain why there was a burst of concerted activity that started in 2019 (sometimes independently across nation-states, although one can never be certain of geopolitical causal factors), and global partnerships and initiatives such as the GPAI started to emerge. However, it bears noting that such initiatives take much agreement and coordination to be instituted and were not created overnight. Hence, international bodies, including within the UN, were likely paying attention to the progress of AI well before then.

---

[2] The other two themes include *future of work*, and *innovation and commercialization*.

[3] United Nations Interregional Crime and Justice Research Institute.

[4] United Nations Educational, Scientific and Cultural Organization.

It is difficult to know what the natural progress of such agencies and initiatives on AI regulation would have been if the COVID-19 pandemic had not occurred. Regardless (and notwithstanding recent crises such as the Ukraine-Russia War, and fears of economic slowdown and stagflation), it is likely the issue will pick up steam again, since AI applications are continuing to be adopted with vigor across industries. Recent troubles with labor shortages, increased wages, stagnant productivity, and supply chains issues[5] may even have accelerated long-run trends toward greater automation. There is already some evidence that the government has turned its attention back to regulating AI. We mentioned the warning issued by a US federal agency on AI screening software potentially discriminating against job candidates (e.g., who have a disability or are otherwise flagged by such systems as being unsuitable for the job), but other instances will also become apparent in a US-based case study that we subsequently discuss.

5.4.1 Case Study 1: The European Union General Data Protection Regulation (GDPR)

In May 2018, data protection reforms that had long been in the planning stage started to be enforced across Europe. These reforms, aptly titled as *General Data Protection Regulation (GDPR)*, have since been in place and serve as both a current international standard and a set of much-needed safeguards for protecting individuals' personal information. Prior to GDPR, there had been a patchwork of data protection rules implemented over two decades (mostly since the advent and uptake of the Web), some dating all the way back to the 1990s. Much has since changed, examples including the emergence of the Big Tech corporations, winner-takes-all market forces in the digital economy, and the ubiquity of the smartphone and social media.

It may seem strange that we would be emphasizing a regulation that seems focused on privacy rights rather than on AI *per se*, but there is a close connection between privacy, data protection, and AI. Modern AI systems, including deep neural networks, are extremely data hungry. Companies, especially Big Tech firms such as Google (Alphabet), Facebook (Meta), and several others, often rely on the data collected from users to power targeted advertisements and recommendations that, in turn, drive their growth, revenue, and profits. Although these firms have attempted to diversify some of their digital offerings, they are still, to a sizable extent, driven by advertising as their proverbial bread-and-butter.

As described in an accessible article [5], *personal data* is at the core of GDPR and is broadly meant to be understood as information that allows a living human being to be identified (directly or identified) from available data. Such information can

---

[5] While these issues have now been largely resolved, they have forced companies to re-consider just-in-time manufacturing and global supply networks and re-locate production to their home countries, with higher labor costs.

be obvious, such as names, locations, or identifiers, but it can also be information less apparent (e.g., IP addresses could be considered as falling in this category). Importantly, "pseudonymized" data can also be considered personal, if deemed sufficiently identifying.

Certain special types of sensitive personal data are accorded greater protections by GDPR. These include information about ethnic or racial origin, religious beliefs, political opinions, trade union membership, biometric and genetic data, data concerning the individual's sexual orientation, and health information.

Although an EU regulation, GDPR could also apply to companies based outside of the region, if the company chooses to do business in the EU, which is usually a given for large companies due to the size of the market and its purchasing power. For example, if a business is based in the United States but collects or uses the data of EU citizens, GDPR would still apply to that business.

Seven key principles guide the interpretation and enforcement of GDPR and are laid out in Article 5 of the legislation. We say "interpretation" because the principles are not "hard" rules: rather, they are meant to be understood as a framework for scaffolding the broad purposes and intent of the regulation. Many of these principles have been heavily influenced by, and significantly overlap with, many of the previous data protection laws that had existed before the GDPR.

These seven principles are *lawfulness, fairness, and transparency; purpose limitation; data minimization; accuracy; storage limitation; integrity and confidentiality (security); and accountability*. Of these, the accountability principle is perhaps the most novel, especially in a country like the UK, where all the other principles had existed in some form under their 1998 Data Protection Act.

Based on the work in [8], below, we briefly summarize these seven principles, which are espoused right at the beginning of the legislation, informing all else that follows. A building block that is therefore fundamental for good data protection practice is compliance with these principles. Article 83(5)(a) of the regulation states that infringements of the principles for processing personal data are subject to the highest tier of administrative fines (which in the UK, for instance, would have amounted to fines as high as 17.5 million pounds, or 4% of the organization's total annual global turnover, whichever is higher).

1. **Lawfulness, fairness, and transparency:** The principle states that personal data shall be processed lawfully, fairly and in a transparent manner in relation to the data subject ('lawfulness, fairness, transparency').
2. **Purpose limitation:** The principle that personal data be collected for specified, explicit, and legitimate purposes, and not be processed further in a manner incompatible with those purposes [Article 5(1)(b), GDPR]. Exceptions include further processing with the data subject's consent, processing on the basis of EU or member state law, or processing for public interest purposes.
3. **Data minimization:** This principle essentially states that organizations should not collect more personal information than they need from their users. The principle is important in an age when we are creating more information than ever. It is designed to preempt organizations from overreaching with the type of data

they collect about people. For instance, it is unlikely that an online retailer would need to collect people's political opinions when they sign-up to the retailer's email mailing list to be notified when sales are taking place.

4. **Accuracy:** This principle states that every reasonable step must be taken to ensure that personal data that is inaccurate, with regard to the purposes for which it is processed, is erased or rectified, without delay.

5. **Storage limitation:** This principle states that personal data be kept in a form which permits identification of data subjects for no longer than is necessary for the purposes for which the personal data are processed. However, personal data may be stored for longer periods insofar as the personal data will be processed solely for archiving purposes in the public interest, scientific or historical research purposes or statistical purposes, in accordance with Article 89(1), and subject to implementation of the appropriate technical and organizational measures required by the regulation in order to safeguard the rights and freedoms of the data subject.

6. **Integrity and confidentiality (security):** This principle states that personal data must be protected against unauthorized or unlawful processing, as well as accidental loss, destruction, or damage. We note that under the UK 1998 Data Protection law, security was the seventh principle outlined, and over twenty years of best practices now exist on implementing this principle. Organizations must take appropriate information security protections to reasonably ensure that information cannot be accessed by hackers, or accidentally leaked due to data breaches or other such events.

   We note that GDPR does not provide specifics about what good security practices look like, in recognition that it may be different for every organization. For example, a bank would be expected to protect information in a more robust way than (say) a local book club. Broadly speaking, proper access controls to information should be put in place, websites should be encrypted, and "pseudonymisation" is encouraged. If a data breach occurs, data protection regulators will carefully consider a company's information security setup when the breach occurred to determine the extent of fines and punishments.

7. **Accountability:** This principle states that organizations must document how personal data is handled and the steps taken to ensure only people who need to access some information are able to. It was added to the GDPR to ensure that companies can prove that they are working to comply with the other principles that constitute the regulation. Accountability can also include training staff in data protection measures, and regularly evaluating data handling processes. For companies with more than 250 employees, there is also a requirement to have documentation on why people's information is being collected and processed, description of the information that is being held, how long said information will be kept for, and details of technical security measures in place. Article 30 states that most organizations must also keep records of said data processing, sharing, and storage.

   We note that the accountability principle is crucial in a defensive sense for the organization, especially if it gets investigated for potentially violating one of

GDPR's principles. Having an accurate record of all systems in place can help such an organization to prove to regulators that it took the GDPR obligations seriously.

5.4.1.1 Enforcement of GDPR

The GDPR is a good example of a regulation that depends more on enforcement than on the actual wording of the law. At the time it was first passed, the jury was still out on whether it would have "teeth," i.e., would regulators fine businesses that do not comply such that it deters them in the future? Deterrence is important because, if the fines are too inconsequential, non-compliance with the regulation practically amounts to an added tax or cost of doing business.

Since the regulation has been passed, many would argue that its enforcement has been reasonably effective, although some believe it would need to become more aggressive to truly deter "bad behavior" (as opposed to becoming a "cost of doing business" that organizations take into account, while continuing to flout the principles). One of the biggest fines under GDPR to date has been against Google. In France, the National Data Protection Commission fined Google more than 50 million Euros for two stated reasons: not providing adequate information to users about how their data (collected from about twenty different services) was getting used, and not getting proper consent for processing those users' data. Beyond Google, a number of (less newsworthy) instances also come to mind. For example, Bulgaria's DSK Bank was fined under GDPR for accidentally disclosing customer details, and fines were also lobbied against the app of La Liga (a professional sports league) for collecting data on people who downloaded it. Nevertheless, the jury is still out on whether its enforcement is truly stringent enough to lead to a change in behavior, or whether companies have just come to see it as another cost.

Within the UK (at least prior to Brexit), the Information Commissioners Office issued a "notice of intent" to two major companies (British Airways and Marriott) for breaching GDPR. Originally, for example, Marriott was going to be fined almost a hundred million pounds, but in November, 2020, it was reported that it had managed to secure a reduction of 80% from the original fine. The details of the case are complicated (we refer the interested reader to [13] for more details), but the key point to note is that the GDPR is not just targeted at Big Tech; furthermore, given that it is still a young regulation, it has already been effective at forcing many companies to comply with its spirit to a pragmatic extent.

5.4.2   Case Study 2: The United States National Defense Authorization Act (NDAA)

In the *Introduction*, we had briefly mentioned the Artificial Intelligence Capabilities and Transparency (AICT) Act and the Artificial Intelligence for the Military (AIM) Act as signature pieces of AI legislation that had been included in the United States National Defense Authorization Act (NDAA) for fiscal year 2022 (signed into law in early 2022 by President Joe Biden). Of these, AICT has a more direct link to AI ethics, while AIM cites as its goal the implementation of "recommendations relating to military training on emerging technologies."

According to [2], the AICT implements the recommendations of the National Security Commission on Artificial Intelligences (NSCAI) final report. The NSCAI NDAA order was established by the Congress through Fiscal Year 2019 to consider the "methods and means necessary to advance the development and improve the government's use of AI and related technology." The AICT is meant to establish a Chief Digital Recruiting Officer within the Department of Defense, the Department of Energy, and the Intelligence Community to identify digital talent needs and recruit personnel, and recommends that the NSF should establish focus areas in AI safety and AI ethics as a part of establishing new, federally funded National Artificial Intelligence Institutes.

Although some press outlets and commentators, such as Fortune [6], opined that the legislation falls "far short" of the calls for regulation that are more consistent with the European Union model (and that scholars in the AI ethics community feel are necessary), it is a much needed first step for a pragmatic AI ethics regulatory regime that could be accepted by US businesses.

At the same time, the 2022 NDAA should not be thought of as any more than a starting point. More focused and operational directives are going to be needed in the future to bake in ethics in AI workflows in the federal government and the Department of Defense. The current legislation aims to direct this responsibility to the director of the National Science Foundation and its AI Research Institutes. Furthermore, commentators have argued that at least three elements need to be borne in mind by any director that best tries to fulfill this role [6]:

1. First, the director should create an AI *use-case archive* or repository. Such an archive could prove instrumental in providing a fine-grained view of each use case currently being deployed (or in various stages of deployment) within the federal government and DoD. It goes without saying that all of the archive will likely not be publicly available; some use cases may even require secret clearance. However, congressional members with the appropriate clearances could still use such an archive to perform oversight functions, ensure that the legislation is being adhered to, and have a degree of transparency that often gets lost in bureaucracy.
2. Second, the director should harmonize and synchronize the different ethics vetting frameworks that already exist. Some of these frameworks use terminology and concepts that are slightly different, although none of them is truly contradictory. An example of such a framework is the comprehensive set of Responsible AI Guidelines that was recently published by the DoD Defense Innovation Unit. Some have argued that this is among the most practicable (and robust) frameworks available for guiding DoD contractors in the implementation and deployment of use cases. Other influential examples that could be considered are the EU Guidelines on Ethics in Artificial Intelligence and the World Economic Forum AI Government Procurement Guidelines.

3. Third, it is extremely important to have a good public communications strategy, especially given that trust has been eroded greatly in establishment institutions, and the government, of late. The strategy should focus on transparency, goals, and use clear language. It must adopt a cautious note and assure citizens that their rights are not being trampled upon, and that this is not a "Big Brother" effort. As the recent demise of a government effort indicates [7], fostering and sustaining public trust and communication are non-trivial in today's partisan climate.

5.5   AI Ethics in Research and Higher Education

Policy, regulation, and legislation have an obvious impact on businesses, especially when they are cross-country regulations in large markets such as the European Union, and cover more than just the military. By the time that an issue has been brought before a regulatory or legislative authority, it may be too late for businesses (especially, small and medium-sized enterprises) to have a meaningful say in the matter. Where, then, can businesses look to in order to stay ahead of the curve?

As with technical research, we suggest that it may be worth the time of business leaders across industries to pay due attention to research emerging from interdisciplinary AI centers at major institutions. Universities have started taking the lead on matters of AI safety, ethics, and policy, and it is likely that the research emerging from influential universities will influence decisions made by lawmaking bodies such as the US Congress.

As an example of such a center, Northwestern University recently launched a research hub for AI safety and equity in collaboration with Underwriters Laboratories. The goal of the center (titled the *Center for Advancing Safety of Machine Intelligence* or CASMI) is to "foster research that integrates safety into AI design and development" [11].

While the above is a good example of a center focused specifically on responsible and ethical AI, other such instances can also be cited and are growing in numbers. One such example is the Institute for Human-Centered AI (HAI) founded in Stanford University as recently as early 2019. Within this center, researchers aim to develop AI that can augment and enhance human productivity, and human quality of life. The center is also a testament to the growing importance of augmented AI, which was covered in depth in the previous chapter. Faculty and staff affiliated with HAI include both cutting-edge scientists, as well as scholars studying social movements, educators looking to develop a new generation of pedagogy and curricula, and even lawyers and artists. Yet another example we cite is the AI Now Institute in New York University (NYU) that was founded in 2017 by Kate Crawford and Meredith Whittaker. The institute has established formal partnerships with NYU Law, NYU for Data Science, Partnership on AI, and others and aims to serve as a hub for emerging, interdisciplinary research that sheds light on AI's social implications.

Beyond the United States, examples include the Institute for Ethical AI & Machine Learning, which is a UK-based research center that carries out highly technical research into processes and frameworks that support the responsible development, deployment, and operation of machine learning systems. They are formed by cross functional teams of volunteers including ML engineers, data scientists, industry experts, policy-makers and professors in STEM, humanities and social sciences. Other examples include the Civic AI Lab (a collaboration between the City of Amsterdam, Vrije Universiteit Amsterdam, and University of Amsterdam), the Wadhwani Institute for Artificial Intelligence in India, the Institute for Ethics in Artificial Intelligence (based in Munich, Germany), the Centre for the Governance of AI (GovAI) at the Future of Humanity Institute (based in the University of Oxford, UK), and the Montreal AI Ethics Institute in Canada, to only name a few.

5.6   Conclusion

A key space to watch in the AI regulatory landscape beyond Europe and the United States is China. As discussed earlier in Chap. 3, Chinese "Big Tech" firms such as Alibaba and Baidu have experienced rapid growth and are dominant players in a country that, by most accounts, has emerged as a twenty-first century economic and geopolitical superpower and a counterpoint (in policy discourse) to Western notions of privacy and security. It is not within the realm of our expertise to comment on the merits and demerits of one or the other, but given the size of the Chinese Big Tech (and the Chinese economy more generally), we note the recent regulatory efforts of the Chinese government in reining in Big Tech for an instructive case study that deserves to be analyzed.

As reported by Nikkei Asia [14], for example, Chinese regulators have implemented a new set of rules where, starting from March 1 of 2022, consumers will have the right to turn off app algorithmic recommendations, and see (and delete) the keywords that these algorithms are using to target them. Similar to Big Tech in the United States, such recommendations are cornerstone elements of large tech companies in China such as Alibaba, ByteDance, and Tencent, since they use such algorithms to predict and personalize what consumers will watch and buy online. Similar to the GDPR in the EU, these regulations are being enforced by the Chinese government to constrain tech's growing power over the population (and especially, the younger segments). These rules are not unexpected; they are based on a draft released in 2021, with the standards created jointly by the Cyberspace Administration of China, the Ministry of Industry and Information Technology, the State Administration for Market Regulation, and the Ministry of Public Security. Within the United States, similar such discussions are occurring, although it is not clear what the constitutional boundaries of such regulation are. We note these examples only to highlight that this is still an emerging area of policy where much can change at any time, and more rapidly than organizations or individuals might expect.

The *long-term* impact of these regulations, and other regulations that may be arising in the near future, remains to be seen. A key element to note here is that concerns about AI are no longer national, or even Western, issues. As the pace of AI innovation continues, and corporations amass ever greater power in regional and world economies, governments, and regulatory bodies are starting to step up to the ambitious task of drawing the boundary lines to protect consumers and other social stakeholders. We suspect that the regulatory landscape will have evolved considerably, and many more lessons will have been learned, even just 3-5 years from the time this book is published. Hence, no company innovating in AI, or wishing to adopt AI innovations, can afford to ignore AI regulation and ethics as just a "Big Tech" (or alternatively, regionally constrained) problem.

References

1. Virtue ethics (2016). URL https://plato.stanford.edu/entries/ethics-virtue/
2. Portman, Heinrich announce bipartisan artificial intelligence bills included in FY 2022 National Defense Authorization Act. (2021). URL https://www.portman.senate.gov/newsroom/press-releases/portman-heinrich-announce-bipartisan-artificial-intelligence-bills-included
3. Angwin, J., Larson, J., Mattu, S., Kirchner, L.: Machine bias. (2016). URL https://www.propublica.org/article/machine-bias-risk-assessments-in-criminal-sentencing
4. Buolamwini, J., Gebru, T.: Gender shades: Intersectional accuracy disparities in commercial gender classification. In: S.A. Friedler, C. Wilson (eds.) Proceedings of the 1st Conference on Fairness, Accountability and Transparency, *Proceedings of Machine Learning Research*, vol. 81, pp. 77–91. PMLR (2018). URL https://proceedings.mlr.press/v81/buolamwini18a.html
5. Burgess, M.: What is GDPR? the summary guide to GDPR compliance in the UK. (2020). URL https://www.wired.co.uk/article/what-is-gdpr-uk-eu-legislation-compliance-summary-fines-2018
6. Griffin, W.: America must win the race for A.I. ethics (2022). URL https://fortune.com/2022/02/15/america-must-win-the-race-for-a-i-ethics-tech-artificial-intelligence-politics-biden-dod-will-griffin/
7. Hart, B.: Poorly conceived Biden disinformation board put on pause. (2022). URL https://nymag.com/intelligencer/2022/05/poorly-conceived-biden-disinformation-board-put-on-pause.html
8. ICO: Guide to the general data protection regulation: The principles (2022). URL https://ico.org.uk/for-organisations/guide-to-data-protection/guide-to-the-general-data-protection-regulation-gdpr/principles/
9. Department of Justice, O.o.P.A.: Justice department and EEOC warn against disability discrimination. (2022). URL https://www.justice.gov/opa/pr/justice-department-and-eeoc-warn-against-disability-discrimination
10. Langston, J.: Who's a CEO? google image results can shift gender biases. (2015). URL https://www.washington.edu/news/2015/04/09/whos-a-ceo-google-image-results-can-shift-gender-biases/
11. Now, N.: Northwestern launches research hub for AI safety, equity. (2022). URL https://news.northwestern.edu/stories/2022/02/northwestern-launches-research-hub-for-ai-safety-equity
12. OECD: OECD AI principles overview (2019). URL https://oecd.ai/en/ai-principles
13. Wayte, D., Connor, J.: Marriott secures 80% reduction in ICO fine, but here's what you missed. (2020). URL https://www.orrick.com/en/Insights/2020/11/Marriott-Secures-80-Reduction-in-ICO-Fine-but-Heres-What-You-Missed
14. Zhihang, D., Yi, D., Han Wei, C.: China tightens grip on big tech's use of algorithms (2022). URL https://asia.nikkei.com/Spotlight/Caixin/China-tightens-grip-on-Big-Tech-s-use-of-algorithms

What Is on the Horizon?

6

6.1 Introduction

This book aimed to clarify the role that AI will play in industries of the future, either as a complement or as a primary technology without which the industry cannot exist. We considered such a role from multiple perspectives, including the proper way in which costs, benefits and return on AI investments and implementations should be considered (Chap. 2), as well as the places today where AI is being applied and which will likely form the cradle for industries of the future (Chap. 3). Although specific companies will go in and out of business as the times change, a small case study of an advanced AI area (neural language models) shows just how far the field has come, and why transformer-based neural networks give us reason to believe that there is a seismic shift in the tasks that AI can now accomplish. We also considered the importance of augmented AI in a practical world where the workforce cannot be ignored (Chap. 4) and concluded in the previous chapter with the role of regulations and governance (Chap. 5).

Although many of the issues we have covered in the previous chapters are already starting to manifest in terms of their impact (such as the very real changes that have been brought about in companies' view of AI regulation due to GDPR), we close the book in this chapter with a short review of issues that we believe have significantly *longer* time horizons. While some have longer time horizons than others, each issue (we believe) will eventually become a concern, if they are not already on the radar of many of the larger companies. Furthermore, although we frame these as *issues*, a more productive approach may be to view them (equivalently) as *opportunities*. We have no doubt that the companies and organizations that use these as tailwinds rather than headwinds, and as impetus for further innovation, will have greater probability of success in the long run over those that do not.

© The Author(s), under exclusive license to Springer Nature Switzerland AG 2023
M. Kejriwal, *Artificial Intelligence for Industries of the Future*, Future of Business and Finance, https://doi.org/10.1007/978-3-031-19039-1_6

6.2   Can AI Copyright Its Own Art?

The US Copyright Office has said that art created by AI cannot be copyrighted. However, the matter was controversial, to say the least. Indeed, even in this decision, the Office had actually rejected an appeal of its previous decision through a three-person board. According to an article in Smithsonian, the copyright application was for an AI-created image called "A Recent Entrance to Paradise." As reported in The Verge, "A Recent Entrance to Paradise" is part of a series that has been described as a "simulated near-death experience" [81]. A deep neural network algorithm re-processes pictures to create images that can only be described as hallucinatory, guided by a fictional narrative about the afterlife. Minimal human intervention was involved in the creation of the reprocessed image, which was the point of contention that came before the Copyright Office.

Both the initial application and appeal are part of a project by legal scholar Stephen Thaler. Thaler is using the application to test what boundaries of intellectual property law are in the United States and other countries in an age when AI seems likely to become a creative force [100]. While, thus far, the United States has ruled on the side that claims that copyright protections can only be held by a human (which is obviously not welcome news to technologists), we suspect the matter is far from settled.

To quote the board in its decision, "the nexus between the human mind and creative expression" is a vital element of copyright. Its perspective is that, while copyright law does not *directly* outline rules for non-humans, the courts have not been accommodating to claims that animals and non-human entities, such as divine beings (and now AI) can take advantage of copyright protections. Concerning the issue of divinity, there is already a decision from the late 1990s wherein a book of "divine revelations" could be protected if human arrangement and curation was involved. Concerning animals, the court has found that a monkey cannot sue for infringement. While this might suggest that the deck is stacked against an AI in that "non-human expression is ineligible for copyright protection" to quote from the board's decision, it does not necessarily imply that *any* art with an AI component would be ineligible.

However, to concede along the lines above would have been incongruent with Thaler's goals, who has sought to establish that machine-created works should receive protection. This is not merely to prevent people from infringing on the work. It is possible that a different decision would be reached in a more realistic situation where a company or individual artist (or inventor) brings the lawsuit that a copyright application is, in fact, an infringement of their copyright (e.g., if someone tried to take an image copyrighted by them and then "trained" a machine to produce something reasonably similar). Alternatively, Thaler himself has the option to bring another lawsuit to challenge this decision.

We should also note that, internationally, the legal situation is even murkier, albeit with some promise (for Thaler). In July of 2021, the Federal Court of Australia handed down a decision (again, in a case filed by Thaler), wherein it permitted the listing of the AI system DABUS[1] as an inventor in a patent application [67]. It is interesting to explore what implications this decision could have in the field of copyright. The case is so named because it refers to a patent application where AI DABUS was listed as an "inventor." While initially rejected by multiple Intellectual Property (IP) protection offices around the Western world, such as the UK, Germany and the United States, the Federal Court of Australia decided to hear the appeal with the finding that the AI *could* be listed as an inventor under the Australian Patent Act [63].

The court found that (compatible with the goal of the act), as the developer and owner of DABUS, Thaler would own the patent. The court felt that such a decision is in the spirit of the act to promote innovation, and that the act does not, either explicitly or through implication, preempt the listing of an AI as an inventor. As with the US case, this decision should not (at the time of writing) be thought of as final, since the decision has already been appealed before the full Federal Court. An outcome is pending when we last checked.

It is evident from this context that the legal issues around AI, and its status as an inventor (or as a creator, in the broader sense of the word) is far from over. AI-specific legal issues of this kind are currently only of theoretical importance in industry, since almost every patent or invention in industry today still involves significant human elements. But given the rapid pace of technological advancement, it is not unlikely that the boundaries will get blurred. At the same time, even without copyright or patenting, there are still protective measures that a company could take (and do take) to ensure that its investment in building such an AI is not wasted. More practically, the fact that there are serious discussions over whether AI can be granted a copyright on its art shows how much of a tool it can be in the creator economy. The issues that we raised in Chap. 4 about augmented AI and its potential are likely just the beginning of what this development means for the workforce.

6.3   Legal Issues Around Deepfakes

Unlike the AI copyright issue, which can be resolved by legal systems around the world in due course without significant urgency (at least at the moment), the same cannot be said for so-called *deepfakes*. The word *deepfake* is a portmanteau of "Deep Learning" and "Fake." Deep learning algorithms, based on Generative Adversarial Networks or GANs [97], can be trained on massive data sets and used to generate fairly convincing, but fake, images (and by extension, videos). They need a prompt of only a few hundred images, a number that keeps coming down, even as the realism of the image continues to increase. At risk of sounding simplistic, these algorithms "swap" a source face with a target face. Of course, if simple "cut-paste" swap was all that was needed, deepfakes would not cause such alarm, and image

---

[1] Device for the Autonomous Bootstrapping of Unified Sentience.

doctoring would be detectable, even just to the human eye. However, deepfakes are so convincing that the swapped images look real.

Malicious use of deepfakes has relatively recent origins. Arguably, the first instance was in 2017, when a software developer nicknamed "deepfakes" swapped Hollywood celebrities faces onto those of porn artists and posted his work on Reddit. The full implications of this creation are still being critiqued and discussed [99]. Since that time, novel applications for deepfake content have also emerged. Because these applications tend to be free, replicable, and easily accessible, they enable many people to create such deepfake content either of themselves or (as in the Reddit case) of well-known public figures, such as politicians, actors, and Hollywood celebrities. Famous examples of public figures of whom deepfakes have been created include Mark Zuckerberg, Donald Trump, and even Salvador Dali [93]. As noted above, the technology is now at a stage where it can be difficult for the layperson to distinguish between the real and the deepfake. This has also opened up a new frontier of AI research into robust deepfake detection [65]. Some fear that this may lead to a new kind of "deepfake" arms race where increasingly sophisticated rounds of deepfake detection and generation technology lead to a zero-sum game where no content can be trusted [11]. Malicious uses are not just limited to porn (including revenge porn) but also could be used to push conspiracy theories and "post-truth" politics and facts, manipulate unsuspecting segments of the public, intervene in elections, and violate intellectual property and personal data protection rights.

However, deepfakes have also found wide and beneficial usage in industries ranging from health to entertainment. In the health sector, for example, they have been used to detect tumors [25]. More than any other modern technology, arguably, deepfakes showcase the truth of the classic statement that technology is not "inherently" good or bad; rather, its usage is.

Regulation is starting to catch up on this matter. We focus on intellectual property as it allows a natural point of comparison with the previous section. In its publication titled "Draft Issues Paper On Intellectual Property Policy And Artificial Intelligence" the World Intellectual Property Organization (WIPO) suggests two important questions of IP rights that need to be addressed in the context of deepfakes. First, since deepfakes are created on the basis of data that may be the subject of copyright, who should own the copyright to the deepfake (or content generated by the deepfake)? Notice the parallels of this question to the issue in the previous section about whether an AI itself can be granted a copyright. Second, should there be (and what would be the nature of) a system of equitable remuneration for individuals whose likenesses and "performances" are used in the deepfake?

Even WIPO concedes that deepfakes may end up causing more severe problems such as violation of privacy and personal data protection rights, than copyright infringements. The main concern in an IP context is whether copyright should even apply to deepfake content. According to WIPO, if deepfake content is subject to copyright, it makes the most sense that the copyright should belong to the inventor of the deepfakes. A parallel here that we draw is photography. Just like the subject of a photograph may not own the copyright to their own photograph; similarly, the subject of deepfakes likely would not own a copyright interest in their own image, especially if generated in a non-malicious and purely creative context. However, personal data protection regulation may still apply, as would local regulation on the misuse of deepfakes. The GDPR, discussed in the previous chapter, states in Article 5 that *personal data shall be accurate and, where necessary, kept up to date; every reasonable step must be taken to ensure that personal data that are inaccurate, having regard to the purposes for which they are processed, are erased or rectified without delay ('accuracy').*

From this perspective, if deepfake content is inaccurate or false, it may be subject to rectification or erasure without delay. Even if it does not fall under that umbrella, the victim may still be able to exercise their right to be forgotten, which is granted to European residents in Article 17 of the GDPR as a *right to erasure*. If Article 17 is exercised, the victim could then have the right to force the "data controller" to erase their personal data without undue delay.

In recognition of deepfakes being a potentially dangerous tool for spreading misinformation and malice, both states and Big Tech are already taking action on their misuse. Big Tech companies have developed and applied some of the tools to detect deepfakes and in states like Virginia, Texas, and California, there is legal regulation of deepfakes. As one example, in Virginia, criminal penalties can be imposed for distributing non-consensual deepfake pornography. Similarly, Texas prohibits the creation and distribution of deepfake videos that are intended to harm candidates for public office or influence elections. We expect that more regulation is on the horizon, and that the landscape will evolve as the technology becomes even more advanced. The industry will also mature in its approach to self-regulation of deepfakes (e.g., it would not be a stretch to imagine that some of the Big Tech platforms may end up banning deepfakes if the problem gets severe). For other industries, the IP and copyright issues may be more pressing in the medium term.

6.4 AI's Explainability Crisis

Despite their impressive gains in accuracy, one criticism that has often been lobbied at deep learning algorithms is that they are akin to "black boxes" in that they are unable to satisfactorily *explain* why they output what they do. For example, if a tweet was labeled as having "negative" sentiment, what is the explanation for that label? Did one or two words make the difference? Was it the tone of the tweet? Unfortunately, without an additional layer of interpretation, the deep learning (by itself) would not be able to answer either one of the two questions above, let alone provide an explanation on its own.

The lack of providing satisfactory[2] explanations is not just an academic concern; businesses are thinking twice about using the technology, especially in areas such as government policy, credit allocation, and healthcare, since there is natural discomfort with just accepting an algorithm's outputs as gospel. Some have suggested that, because of the pervasiveness of deep learning algorithms, the field is facing an "explainability crisis" of sorts.

Others see a business opportunity in the crisis, especially in healthcare. In the last few years, several companies have emerged selling software or solutions that purport to offer explanatory insight into an AI's decision making. Such software, when they work properly, are especially useful when interpreting outputs obtained for medical images and multi-modal data. We provided an important case study in radiology earlier (Chap. 4). Unfortunately, there are many instances where such methods turn out to be faulty or inadequate, as recent research has shown.

The problem is well recognized in the academic community. Numerous explainable AI algorithms have been proposed, including LIME [105], the use of Shapley Values [88] and GradCam [84]. While the technical details behind these are beyond the scope of this book, the intuition behind some of these methods is that they operate as a sort of "counterfactual" whereby they change some datapoints that are input until the algorithm is found to change its prediction. The premise is that the datapoints that were changed must be important enough for the final label, and hence, they can be used to suggest a possible explanation for the predicted label.

An immediate problem with this method can be understood in the context of generating satisfactory explanations. Even if a feature, or set of features, is deemed to be important, the first question is whether they provide a sound or complete interpretation in themselves. In other words, they may not constitute a *unique* solution. Another problem is if they yield a counter-intuitive explanation, since at least two possibilities arise: either the algorithm is wrong or it has discovered a signal that the doctor or the medical field may not have considered as a cause but is nevertheless significant. The worry is that it is difficult to distinguish between the two possibilities, leading to further erosion of trust in the AI.

The methods described thus far are commonly known as "post-hoc" methods since they attempt to explain the AI's outputs after the fact. Such methods can be thought of as reactive in scope. An alternative line of thought considers explanations as a first-class citizen by being more proactive in the training process itself. For example, during training, the AI could be optimized so that it is not just outputting a label but also needs to identify (say) "prototypical" features in an image that could serve as a natural intermediate explanation for the final output. However, research looking at multiple explainability algorithms found that they also rest

---

[2] The word "satisfactory" is important and often implicit when explanation in AI is being discussed. One could always make the trivial claim that a deep learning system responded the way it did because neuron X fired, or the cross-entropy loss was Y, given the output, etc. These are only mechanistic claims, however, and not the kind of "semantic" explanation that gets to the heart of the matter (and provide necessary context to humans).

heavily on human interpretation. Were the features detected with sufficient accuracy, and appropriately weighted to yield the final output, for instance?

An even more worrying problem that was only documented recently by a multi-institutional group of researchers is that state-of-the-art explainability algorithms often disagree on the explanation for an algorithm's outputs [28]. This is somewhat like the "unique solution" problem we mentioned earlier, only much worse. The researchers found that, in real-world settings, people often had no way of resolving these differences, and that the explanation could reinforce pre-existing ideas, a form of "confirmation bias", which has been a well-documented problem in the decision science literature.

We end this section with the somewhat non-obvious conclusion by the authors in [28], whereby it is argued that explanation may not be the correct area of focus, especially in settings (such as medical diagnosis) where confirmation bias is a potential problem. Instead, the authors argue that the correct requirement for using or trusting the algorithm is that its performance has been tested in a rigorous and scientific manner. Interestingly, explanation is not the standard requirement when testing or administering a drug (for example). Once the safety of the drug has been established, a clinical trial needs only to establish the efficacy of the drug against a baseline (such as a placebo, or in some cases, the prior standard of treatment). Explaining why the drug worked, or did not work, is not the guiding concern. In other words, as long as efficacy has been established through rigorous scientific and statistical standards (e.g., by enforcing minimal probability of Type I error through high statistical significance, and high enough statistical power through mechanisms such as large sample sizes), approval can be obtained through the relevant government agency.

Considering that this is the prevailing standard in other areas of medicine, it seems fair to ask why a different standard should apply to algorithms. The issue, however, is far from settled. We also note that, while this issue is of particular concern to the healthcare industry, it will increase in importance as more decision making power is relegated to algorithms. There may well be an entire industry of the future that dabbles in this specific problem area, which has already spawned a cottage industry of sorts in the startup space.

6.5   More Vigorous Algorithmic Regulation

One of the indirect messages of this book is that no industry today (and even more so, in the future) seriously looking to use AI can ignore issues of regulation and governance. Deepfakes were an example of how powerful AI technology can be misused, and certain nefarious uses of which are already outlawed. Copyrighting and patenting AI creations falls more within the legal gray-zone. What about algorithms in the here and now, including in important socio-economic domains such as finance, criminal justice, and data collection (and sale) from mobile apps?

Already, the Federal Trade Commission (FTC) in the United States is requiring companies to delete algorithms that have been trained on data that violate data privacy standards [5]. As noted by the tech publication Protocol, the FTC has recently reached at least three high-profile settlements with companies for violating such standards and regulations. Some of these cases involved illicitly obtained data. A particularly recent example is a settlement reached in early March (2022) with WW International (formerly known as Weight Watchers). The company had used a healthy eating app that collected data on children that were as young as eight, but without first obtaining parental approval. Experts seem to agree that this is a good move by the government to preempt (or at minimum, severely discourage) companies from developing or training algorithms that benefit from data obtained in a questionable manner.

It is also possible to interpret "algorithmic regulation" from a completely different angle: can regulation be more streamlined by *using* algorithms? While governments have voiced support for the notion that algorithms could be used to inform decision making in policy (e.g., to optimize resource allocation), there are also concerns that such algorithmic regulation could have unintended consequences, especially for those already disenfranchised in our society. A case study is the ongoing controversy around "predictive policing", which some have argued exacerbates racial disparity and unfairly (and excessively) targets minority neighborhoods. In general, civil rights advocates have voiced concerns about using algorithms in policing and criminal justice, and about the importance of holding these algorithms accountable. Thus far, it is not clear that algorithmic accountability has been secured in these societally important spheres of public life. On the technical front, it is even less clear that existing algorithms can be audited in a way that ensures fairness, although much research on this topic has been forthcoming, especially recently [73].

There is little doubt, however, that we do not have the luxury of time for resolving these thorny issues, as algorithms continue to become more pervasive in political, economic, and social life. The field of "regulation scholarship", increasingly intertwined with policy, has been demonstrating ever greater interest in exploring the full implications of algorithms both for and *in* regulation. We point the interested reader to an influential workshop on this matter in 2017 appropriately titled "Algorithmic Regulation" [3], as well as a more recent book on the subject [102], and an earlier research article on algorithmic regulation and the rule of law [31]. Needless to say, many open and controversial questions remain, some of which are reviewed in the cited article. Industries of the future, including current efforts in Big Tech and startups, cannot ignore these questions and leave them to scholarly and governmental efforts, as their effects will be felt on the companies. This may explain why all the Big Tech companies are known for engaging in extensive lobbying activities [27, 90]. Whether considered as a positive or a negative, such activities have come to constitute an increasingly important element of doing business in today's regulatory environment, especially as AI becomes a more prominent topic in public discourse.

6.6   Increasing Convergence of Emerging Technologies

In the first chapter, we had briefly described some of the other "emerging technologies" that are generally thought of as distinct from AI. These technologies include blockchain and quantum computing. As these technologies, some in different stages of maturity than others, advance, there is reason to believe that they will start converging in real applications, especially those in support of *human-machine teaming* . In defense industries, there is considerable interest in such a concept. Lockheed Martin, for example, states in a video that "human-machine teaming represents the future" [1]. They re-visit the theme that we had presented as augmented AI in Chap. 4, but through the lens of defense. AI is cited as adding value to almost every product and system, including both "military and commercial customers." This is direct evidence of the positive systemic role that AI can play. Autonomous systems with a significant AI component will not only change the way that the military operates its forces but also firefighting, space exploration, and exploring and mapping the depths of our oceans.

Quantum computing presents a good case study of how AI can converge with other emerging technologies, while still maintaining a distinct flavor of its own. The nascent field of *quantum AI* is practically understood today as the use of quantum computing for more powerful machine learning (ML) [23]. There is no dispute today that ML is compute-heavy, but that the computations have a specificity that can be exploited for greater speed [103]. This largely accounts for the success of graphics processing units (GPUs), as well as more recently invented tensor processing units (TPUs) that were developed by Google and are available both on the cloud (and for sale in smaller versions) [35]. However, quantum AI is fundamentally different from these in that it involves both hardware innovation (where core quantum computing research plays the leading role) and a fundamentally different way of designing algorithms such that "qubits" become a first-class citizen [79].

Is quantum AI a purely theoretical promise at the moment? We would argue that, in fact, considerable practical progress is already being observed. Google announced an open-source library for quantum ML called TensorFlow Quantum (TFQ) [7], building off of its successful TensorFlow package for working with neural networks. This library was developed in collaboration with Volkswagen, University of Waterloo and X (Google's "moonshot projects" company). It demonstrates the beginnings of both technological and industry convergence. While the future is still uncertain, we hypothesize that such collaborations (between companies in multiple industries, as well as academia) will only become more commonplace, as the promises and challenges of emerging technologies expand. Quantum AI promises to revolutionize search, decision problems, classical learning and optimization, and even game theory, which continues to be important in social sciences and economics. However, critical milestones must be achieved (in a cost-efficient manner) before such promises can be realized. A full discussion of quantum AI is not within the scope of this book, but we cite two references for the interested reader [21, 74].

Beyond emerging technologies, there is also hope that AI could be merged as an emerging technology to disrupt long-standing industries such as education. Education has come under increasing scrutiny and criticism lately, not only at the higher-ed level (with ballooning costs and rise in student loans) but also at the K-12 level. Achievement scores in the United States have been falling compared to other developed nations, a cause for national concern. The COVID-19 pandemic made issues worse, with some pessimistic estimates projecting that it may have led to a "lost generation" of students that will take concerted effort to correct [85]. Others are more optimistic about the long-term outcome, but there is no dispute that considerable damage has been done to young students between grades 3–8 [59]. Causes are more controversial (including the role that ineffective remote learning had to play), but there is consensus on the immediate (and negative) short-term impacts of the pandemic on educational achievement of young children [60].

In the long run, AI may be a useful capital technology to address chronic problems in education, including the quality of personalized education delivered to students, staff shortages and burnout [16], and accommodation of different learning styles. However, this is still far from the future. The current record of AI-enabled digital tutors has fallen far short of (arguably, hyped up) expectations. This is not to say that startups and other companies are not trying. For example, a Canadian company called Korbit developed an AI-enabled online tutoring software called Mila that was mentioned in a Fortune newsletter and that was shown in a study [87] (conducted by Korbit, but also the Quebec Artificial Intelligence Institute, and the University of Bath, in England) to lead to higher massive open online course (MOOC) completion rates by software programmers at a Vietnamese tech company. These engineers, who had to learn about linear regression, were divided into three groups, two of which, respectively, took a MOOC, and got individualized feedback from an AI-enabled tutor. The third "control" group used Korbit, but with the tutor functionality disabled. The study found that the engineers in the group with the tutor enabled have higher course completion rates and also higher test scores (by margins of more than 2–2.5 times compared to both of the other groups).

While the study would need to be replicated in other contexts, and with other groups (including younger students in the K-12 system, rather than engineers who have a natural incentive to engage in learning), these results do provide some preliminary evidence that AI-assisted tutoring could have beneficial impacts, either on motivation, efficacy of learning, or both. The real problem with education is the scaling of the interactive human element, rather than delivery of a platform or educational curriculum, which MOOCs have already solved. Because of teacher shortages [16], and wide disparity in the quality of teaching and schools, this element has proven difficult to reliably replicate and deliver at scale to students, especially online. The online learning experience during COVID-19 does not add confidence that such a system would effectively work. It is also unclear if systems like Korbit are superior to such human tutors, even on average. Hence, much work still remains to be done, despite the early promise of some studies.

6.7   Concluding Notes

In this chapter, we provided a small sample of issues that should be on the radar of every industry looking to implement AI now, or in the future. Legal issues are not only the most slow-moving but also the most consequential. The day may not be far when regulation of deepfakes rises to the same level as that currently in place for bioweapons or other technologies that directly affect national security [32]. Deepfakes are already being used for political propaganda, sometimes by foreign actors. On a less urgent note, transformer-based AIs and GANs are proving to have creative powers, and the debate is ongoing on whether their "originality" should be recognized through patents and copyrights. At the same time, we would be remiss not to be optimistic about novel applications and uptake of computer vision that we hear of every few months. We previously discussed how drones and computer vision can help wean us toward clean energy (Chap. 3) by making maintenance of solar farms more efficient. Other applications of computer vision that had long been in the making, such as digitization of census records [17–20, 75], have now become much more feasible in practice because of the enhanced accuracy that modern deep neural networks offer [96, 98].

Regulation *of* algorithms, as well as regulation *by* algorithms, are both pre-eminent research areas of AI governance and policy in the current decade. On the one hand, as argued also in Kahneman's recent book [36], the use of algorithms may help us to reduce system noise and achieve consistency. On the other hand, many people feel a loss of agency, or even feel de-humanized [91], if all decisions were entirely left to an algorithm. If the algorithm were to cause damage, who would be liable? Such a discussion would begin in territory that is familiar to decision-makers; namely through a recognition of risks, costs, and benefits. We are only in the nascent stages of fully understanding such risks, however, and the matter of public trust in AI can never be taken lightly [10, 58, 95].

On a more optimistic note, there is increasing convergence of emerging technologies, including quantum computing, blockchain, and AI. Among other novel applications, AI is also being applied for chip design and optimization of compilers [72, 106], with deep neural networks leading to these becoming more pragmatic possibilities [14, 62]. There is also an active and robust movement on promoting *AI for social good* involving academic, consulting and enterprise stakeholders alike [13, 26, 34, 39, 41, 48, 76, 94]. From a scientific standpoint, the advent of cloud computing, cheap storage, as well as ease of using open-source AI and statistical packages, have led to a resurgence in the "Computational [x]" movement, where traditional fields (such as social science [24,33,56,61,69], medicine [80,82,92,101], and even history [4, 68]) are now being viewed through a computational lens[3] to derive novel insights. Some of these "Computational [x]" fields are themselves

---

[3] It would be simplistic to state this as meaning that computation is merely being applied to these fields. Using Big Data and AI for social science (computational social science), for instance, involves far more than that, including careful methodological planning [2].

starting to intersect to create new lines of inquiry. For instance, computational social science methods are now being applied to understand trends on the internet, including organizational usage of Web resources such as *schema.org* [44, 45, 71, 77, 83], and quantification of bias, political misinformation and sensationalism in Web-based text data sources [89, 104], including, but not necessarily limited to, social media [22]. It has also started intersecting with computational epidemiology [66], with a number of papers on COVID-19 vaccine hesitancy published recently in the social media analytics community [64, 70].

While exciting, we believe this is only the beginning. Scalable and low-code (or even no-code) techniques for programming AI continues to be an important area of research that needs more developmental effort [6, 9, 40, 53], as are tools for visualizing machine learning outputs [47, 57]. At the same time, as we have already noted several times in this book, industry is continuing to invest in, and adopt, AI and advanced computational technologies for a range of creative applications with each passing year. This is also true for government agencies, such as the US Department of Defense. In turn, this provides added impetus for researchers to release tools and algorithms that, while not necessarily the most cutting-edge in terms of their accuracy, are scalable[4] [49] and applied in nature.

Finally, within both the AI and human-computer interaction communities, there is increasing focus on problems of "general" AI [29, 86], rather than narrower problems that can be solved by deep learning, given enough data. Solving AGI would likely require solving several "moonshot" problems, including machine commonsense [43], open-world learning [30, 55], greater explainability from deep learning architectures, context-rich AI [38], and causal reasoning [78], neuro-symbolic reasoning [15, 52], including synergies between deep learning and symbolic knowledge [8, 12]. Whether such general AI will ever be achieved is a matter of great speculation, but one that would have the potential to automate much more than is presently possible, and forever change not only the future of work but also the pace and productivity of industrial progress.

References

1. Optimizing the human-machine team. URL https://www.lockheedmartin.com/en-us/capabilities/autonomous-unmanned-systems.html
2. Alvarez, R.M.: Computational social science. Cambridge University Press (2016)
3. Andrews, L., Benbouzid, B., Brice, J., Bygrave, L.A., Demortain, D., Griffiths, A., Lodge, M., Mennicken, A., Yeung, K.: Algorithmic regulation (2017)
4. Au Yeung, C.m., Jatowt, A.: Studying how the past is remembered: towards computational history through large scale text mining. In: Proceedings of the 20th ACM international conference on Information and knowledge management, pp. 1231–1240 (2011)

---

[4] As case in point, within our own group, researchers have published work that relies on fairly simple algorithms but that are extremely scalable [37, 50], unsupervised [42], easy-to-use (even for difficult problems [46, 54]), and designed to operate on heterogeneous (typically, graph-based) data [51].

5. Berger, C.W.: Privacy law for business decision-makers in the united states. In: Ethics of Data and Analytics, pp. 129–137. Auerbach Publications (2022)
6. Bock, A.C., Frank, U.: Low-code platform. Business & Information Systems Engineering **63**(6), 733–740 (2021)
7. Broughton, M., Verdon, G., McCourt, T., Martinez, A.J., Yoo, J.H., Isakov, S.V., Massey, P., Halavati, R., Niu, M.Y., Zlokapa, A., et al.: TensorFlow quantum: A software framework for quantum machine learning. arXiv preprint arXiv:2003.02989 (2020)
8. Capadisli, S., Cotton, F., Dong, X.L., Guha, R.V., Haller, A., Hitzler, P., Kalampokis, E., Kejriwal, M., Lécué, F., Sivakumar, D., Szekely, P.A., Troncy, R., Witbrock, M.J. (eds.): Joint Proceedings of the International Workshops on Hybrid Statistical Semantic Understanding and Emerging Semantics, and Semantic Statistics co-located with 16th International Semantic Web Conference, HybridSemStats@ISWC 2017, Vienna, Austria October 22nd, 2017, CEUR Workshop Proceedings. CEUR-WS.org (2017). URL http://ceur-ws.org/Vol-1923
9. Chang, Y.H., Ko, C.B.: A study on the design of low-code and no code platform for mobile application development. International journal of advanced smart convergence **6**(4), 50–55 (2017)
10. Chen, Y.N.K., Wen, C.H.R.: Impacts of attitudes toward government and corporations on public trust in artificial intelligence. Communication Studies **72**(1), 115–131 (2021)
11. Chesney, R., Citron, D.: Deepfakes and the new disinformation war: The coming age of post-truth geopolitics. Foreign Aff. **98**, 147 (2019)
12. Cochez, M., Declerck, T., de Melo, G., Anke, L.E., Fetahu, B., Gromann, D., Kejriwal, M., Koutraki, M., Lécué, F., Palumbo, E., Sack, H. (eds.): Proceedings of the First Workshop on Deep Learning for Knowledge Graphs and Semantic Technologies (DL4KGS) co-located with the 15th Extended Semantic Web Conference (ESWC 2018), Heraklion, Crete, Greece, June 4, 2018, *CEUR Workshop Proceedings*, vol. 2106. CEUR-WS.org (2018). URL http://ceur-ws.org/Vol-2106
13. Cowls, J., King, T., Taddeo, M., Floridi, L.: Designing ai for social good: Seven essential factors. Available at SSRN 3388669 (2019)
14. Cummins, C., Petoumenos, P., Wang, Z., Leather, H.: End-to-end deep learning of optimization heuristics. In: 2017 26th International Conference on Parallel Architectures and Compilation Techniques (PACT), pp. 219–232. IEEE (2017)
15. De Raedt, L., Manhaeve, R., Dumancic, S., Demeester, T., Kimmig, A.: Neuro-symbolic= neural+ logical+ probabilistic. In: NeSy'19@ IJCAI, the 14th International Workshop on Neural-Symbolic Learning and Reasoning (2019)
16. Dee, T.S., Goldhaber, D.: Understanding and addressing teacher shortages in the united states. The Hamilton Project **5**, 1–28 (2017)
17. Diesendruck, L., Marini, L., Kooper, R., Kejriwal, M., McHenry, K.: Abstract: Digitization and search: A non-traditional use of HPC. In: 2012 SC Companion: High Performance Computing, Networking Storage and Analysis, Salt Lake City, UT, USA, November 10–16, 2012, pp. 1460–1461. IEEE Computer Society (2012). DOI URL https://doi.org/10.1109/SC.Companion.2012.259
18. Diesendruck, L., Marini, L., Kooper, R., Kejriwal, M., McHenry, K.: Digitization and search: A non-traditional use of HPC. In: 8th IEEE International Conference on E-Science, e-Science 2012, Chicago, IL, USA, October 8-12, 2012, pp. 1–6. IEEE Computer Society (2012). DOI URL https://doi.org/10.1109/eScience.2012.6404445
19. Diesendruck, L., Marini, L., Kooper, R., Kejriwal, M., McHenry, K.: A framework to access handwritten information within large digitized paper collections. In: 8th IEEE International Conference on E-Science, e-Science 2012, Chicago, IL, USA, October 8–12, 2012, pp. 1–10. IEEE Computer Society (2012). DOI URL https://doi.org/10.1109/eScience.2012.6404434
20. Diesendruck, L., Marini, L., Kooper, R., Kejriwal, M., McHenry, K.: Poster: Digitization and search: A non-traditional use of HPC. In: 2012 SC Companion: High Performance Computing, Networking Storage and Analysis, Salt Lake City, UT, USA, November 10–16, 2012, p. 1462. IEEE Computer Society (2012). DOI URL https://doi.org/10.1109/SC.Companion.2012.260

21. Dilmegani, C.: In-depth guide to quantum artificial intelligence in 2022. (2022). URL https://research.aimultiple.com/quantum-ai/
22. Ding, J., Kejriwal, M.: An experimental study of the effects of position bias on emotion cause extraction. CoRR **abs/2007.15066** (2020). URL https://arxiv.org/abs/2007.15066
23. Dunjko, V., Briegel, H.J.: Machine learning & artificial intelligence in the quantum domain: a review of recent progress. Reports on Progress in Physics **81**(7), 074001 (2018)
24. Edelmann, A., Wolff, T., Montagne, D., Bail, C.A.: Computational social science and sociology. Annual Review of Sociology **46**(1), 61 (2020)
25. Falahkheirkhah, K., Tiwari, S., Yeh, K., Gupta, S., Herrera-Hernandez, L., McCarthy, M.R., Jimenez, R.E., Cheville, J.C., Bhargava, R.: Deepfake histological images for enhancing digital pathology. arXiv preprint arXiv:2206.08308 (2022)
26. Floridi, L., Cowls, J., King, T.C., Taddeo, M.: How to design AI for social good: seven essential factors. Science and Engineering Ethics **26**(3), 1771–1796 (2020)
27. Foroohar, R.: Don't Be Evil: The Case Against Big Tech. Penguin UK (2019). ISBN 9781984824004
28. Ghassemi, M., Oakden-Rayner, L., Beam, A.L.: The false hope of current approaches to explainable artificial intelligence in health care. The Lancet Digital Health **3**(11), e745–e750 (2021)
29. Grudin, J., Jacques, R.: Chatbots, humbots, and the quest for artificial general intelligence. In: Proceedings of the 2019 CHI Conference on Human Factors in Computing Systems, pp. 1–11 (2019)
30. Haliem, M., Bonjour, T., Alsalem, A.O., Thomas, S., Li, H., Aggarwal, V., Bhargava, B.K., Kejriwal, M.: Learning monopoly gameplay: A hybrid model-free deep reinforcement learning and imitation learning approach. CoRR **abs/2103.00683** (2021). URL https://arxiv.org/abs/2103.00683
31. Hildebrandt, M.: Algorithmic regulation and the rule of law. Philosophical Transactions of the Royal Society A: Mathematical, Physical and Engineering Sciences **376**(2128), 20170355 (2018)
32. House, W., et al.: National biodefense strategy. e-artnow (2020)
33. Hu, M., Rao, A., Kejriwal, M., Lerman, K.: Socioeconomic correlates of anti-science attitudes in the US. Future Internet **13**(6), 160 (2021). DOI URL https://doi.org/10.3390/fi13060160
34. Institute, M.G.: Applying artificial intelligence for social good (2018). URL https://www.mckinsey.com/featured-insights/artificial-intelligence/applying-artificial-intelligence-for-social-good
35. Jouppi, N., Young, C., Patil, N., Patterson, D.: Motivation for and evaluation of the first tensor processing unit. IEEE Micro **38**(3), 10–19 (2018)
36. Kahneman, D., Sibony, O., Sunstein, C.R.: Noise: A flaw in human judgment. Little, Brown (2021)
37. Kejriwal, M.: Sorted neighborhood for the semantic web. In: B. Bonet, S. Koenig (eds.) Proceedings of the Twenty-Ninth AAAI Conference on Artificial Intelligence, January 25–30, 2015, Austin, Texas, USA, pp. 4174–4175. AAAI Press (2015). URL http://www.aaai.org/ocs/index.php/AAAI/AAAI15/paper/view/9295
38. Kejriwal, M.: Essential features in a theory of context for enabling artificial general intelligence. Applied Sciences **11**(24), 11991 (2021)
39. Kejriwal, M.: Link prediction between structured geopolitical events: Models and experiments. Frontiers Big Data **4**, 779792 (2021). DOI URL https://doi.org/10.3389/fdata.2021.779792
40. Kejriwal, M.: A meta-engine for building domain-specific search engines. Softw. Impacts **7**, 100052 (2021). DOI URL https://doi.org/10.1016/j.simpa.2020.100052
41. Kejriwal, M., Fang, G., Zhou, Y.: A feasibility study of open-source sentiment analysis and text classification systems on disaster-specific social media data. In: IEEE Symposium Series on Computational Intelligence, SSCI 2021, Orlando, FL, USA, December 5–7, 2021, pp. 1–8. IEEE (2021). DOI URL https://doi.org/10.1109/SSCI50451.2021.9660089

42. Kejriwal, M., Miranker, D.P.: An unsupervised instance matcher for schema-free RDF data. J. Web Semant. **35**, 102–123 (2015). DOI URL https://doi.org/10.1016/j.websem.2015.07.002

43. Kejriwal, M., Santos, H., Mulvehill, A.M., McGuinness, D.L.: Designing a strong test for measuring true common-sense reasoning. Nature Machine Intelligence **4**(4), 318–322 (2022)

44. Kejriwal, M., Selvam, R.K., Ni, C., Torzec, N.: Locally constructing product taxonomies from scratch using representation learning. In: M. Atzmüller, M. Coscia, R. Missaoui (eds.) IEEE/ACM International Conference on Advances in Social Networks Analysis and Mining, ASONAM 2020, The Hague, Netherlands, December 7–10, 2020, pp. 507–514. IEEE (2020). DOI URL https://doi.org/10.1109/ASONAM49781.2020.9381320

45. Kejriwal, M., Selvam, R.K., Ni, C., Torzec, N.: Empirical best practices on using product-specific schema.org. In: Thirty-Fifth AAAI Conference on Artificial Intelligence, AAAI 2021, Thirty-Third Conference on Innovative Applications of Artificial Intelligence, IAAI 2021, The Eleventh Symposium on Educational Advances in Artificial Intelligence, EAAI 2021, Virtual Event, February 2–9, 2021, pp. 15452–15457. AAAI Press (2021). URL https://ojs.aaai.org/index.php/AAAI/article/view/17816

46. Kejriwal, M., Shao, R., Szekely, P.A.: Expert-guided entity extraction using expressive rules. In: B. Piwowarski, M. Chevalier, É. Gaussier, Y. Maarek, J. Nie, F. Scholer (eds.) Proceedings of the 42nd International ACM SIGIR Conference on Research and Development in Information Retrieval, SIGIR 2019, Paris, France, July 21–25, 2019, pp. 1353–1356. ACM (2019). DOI URL https://doi.org/10.1145/3331184.3331392

47. Kejriwal, M., Shen, K.: Unsupervised real-time induction and interactive visualization of taxonomies over domain-specific concepts. In: M. Coscia, A. Cuzzocrea, K. Shu, R. Klamma, S. O'Halloran, J.G. Rokne (eds.) ASONAM '21: International Conference on Advances in Social Networks Analysis and Mining, Virtual Event, The Netherlands, November 8 - 11, 2021, pp. 301–304. ACM (2021). DOI URL https://doi.org/10.1145/3487351.3489481

48. Kejriwal, M., Szekely, P.A.: An investigative search engine for the human trafficking domain. In: C. d'Amato, M. Fernández, V.A.M. Tamma, F. Lécué, P. Cudré-Mauroux, J.F. Sequeda, C. Lange, J. Heflin (eds.) The Semantic Web - ISWC 2017 - 16th International Semantic Web Conference, Vienna, Austria, October 21–25, 2017, Proceedings, Part II, *Lecture Notes in Computer Science*, vol. 10588, pp. 247–262. Springer (2017). DOI URL https://doi.org/10.1007/978-3-319-68204-4_25

49. Kejriwal, M., Szekely, P.A.: Scalable generation of type embeddings using the ABox. Open J. Semantic Web **4**(1), 20–34 (2017). URL https://www.ronpub.com/ojsw/OJSW_2017v4i1n02_Kejriwal.html

50. Kejriwal, M., Szekely, P.A.: Supervised typing of big graphs using semantic embeddings. CoRR **abs/1703.07805** (2017). URL http://arxiv.org/abs/1703.07805

51. Kejriwal, M., Szekely, P.A.: Supervised typing of big graphs using semantic embeddings. In: S. Groppe, L. Gruenwald (eds.) Proceedings of The International Workshop on Semantic Big Data, SBD@SIGMOD 2017, Chicago, IL, USA, May 19, 2017, pp. 3:1–3:6. ACM (2017). DOI URL https://doi.org/10.1145/3066911.3066918

52. Kejriwal, M., Szekely, P.A.: Co-lod: Continuous space linked open data. In: M.C. Suárez-Figueroa, G. Cheng, A.L. Gentile, C. Guéret, C.M. Keet, A. Bernstein (eds.) Proceedings of the ISWC 2019 Satellite Tracks (Posters & Demonstrations, Industry, and Outrageous Ideas) co-located with 18th International Semantic Web Conference (ISWC 2019), Auckland, New Zealand, October 26–30, 2019, *CEUR Workshop Proceedings*, vol. 2456, pp. 333–337. CEUR-WS.org (2019). URL http://ceur-ws.org/Vol-2456/paper94.pdf

53. Kejriwal, M., Szekely, P.A.: myDIG: Personalized illicit domain-specific knowledge discovery with no programming. Future Internet **11**(3), 59 (2019). DOI URL https://doi.org/10.3390/fi11030059

54. Kejriwal, M., Szekely, P.A., Knoblock, C.A.: Investigative knowledge discovery for combating illicit activities. IEEE Intell. Syst. **33**(1), 53–63 (2018). DOI URL https://doi.org/10.1109/MIS.2018.111144556

55. Kejriwal, M., Thomas, S.: A multi-agent simulator for generating novelty in monopoly. Simul. Model. Pract. Theory **112**, 102364 (2021). DOI URL https://doi.org/10.1016/j.simpat.2021.102364

56. Kejriwal, M., Wang, Q., Li, H., Wang, L.: An empirical study of emoji usage on twitter in linguistic and national contexts. Online Soc. Networks Media **24**, 100149 (2021). DOI URL https://doi.org/10.1016/j.osnem.2021.100149

57. Kejriwal, M., Zhou, P.: SAVIZ: interactive exploration and visualization of situation labeling classifiers over crisis social media data. In: F. Spezzano, W. Chen, X. Xiao (eds.) ASONAM '19: International Conference on Advances in Social Networks Analysis and Mining, Vancouver, British Columbia, Canada, 27–30 August, 2019, pp. 705–708. ACM (2019). DOI URL https://doi.org/10.1145/3341161.3343703

58. Knowles, B., Richards, J.T.: The sanction of authority: Promoting public trust in ai. In: Proceedings of the 2021 ACM Conference on Fairness, Accountability, and Transparency, pp. 262–271 (2021)

59. Kuhfeld, M., Soland, J., Lewis, K.: Test score patterns across three covid-19-impacted school years (2022)

60. Kuhfeld, M., Soland, J., Lewis, K., Morton, E.: The pandemic has had devastating impacts on learning. What will it take to help students catch up? (2022). URL https://www.brookings.edu/blog/brown-center-chalkboard/2022/03/03/the-pandemic-has-had-devastating-impacts-on-learning-what-will-it-take-to-help-students-catch-up/

61. Lazer, D., Pentland, A., Adamic, L., Aral, S., Barabasi, A.L., Brewer, D., Christakis, N., Contractor, N., Fowler, J., Gutmann, M., et al.: Social science. computational social science. Science (New York, NY) **323**(5915), 721–723 (2009)

62. Li, M., Liu, Y., Liu, X., Sun, Q., You, X., Yang, H., Luan, Z., Gan, L., Yang, G., Qian, D.: The deep learning compiler: A comprehensive survey. IEEE Transactions on Parallel and Distributed Systems **32**(3), 708–727 (2020)

63. Li, N., Koay, T.: Artificial intelligence and inventorship: an Australian perspective. Journal of Intellectual Property Law & Practice (2020)

64. Luo, Y., Kejriwal, M.: Understanding COVID-19 vaccine reaction through comparative analysis on twitter. CoRR **abs/2111.05823** (2021). URL https://arxiv.org/abs/2111.05823

65. Lyu, S.: Deepfake detection: Current challenges and next steps. In: 2020 IEEE international conference on multimedia & expo workshops (ICMEW), pp. 1–6. IEEE (2020)

66. Marathe, M., Vullikanti, A.K.S.: Computational epidemiology. Communications of the ACM **56**(7), 88–96 (2013)

67. Matulionyte, R.: AI as an inventor: Has the federal court of Australia erred in DABUS? Available at SSRN 3974219 (2021)

68. McEnery, A., Baker, H.: Corpus linguistics and 17th-century prostitution: Computational linguistics and history. Bloomsbury Academic (2016)

69. Melotte, S., Kejriwal, M.: A geo-tagged COVID-19 twitter dataset for 10 north American metropolitan areas over a 255-day period. Data **6**(6), 64 (2021). DOI URL https://doi.org/10.3390/data6060064

70. Melotte, S., Kejriwal, M.: Predicting zip code-level vaccine hesitancy in US metropolitan areas using machine learning models on public tweets. CoRR **abs/2108.01699** (2021). URL https://arxiv.org/abs/2108.01699

71. Mika, P.: On schema. org and why it matters for the web. IEEE Internet Computing **19**(4), 52–55 (2015)

72. Mirhoseini, A., Goldie, A., Yazgan, M., Jiang, J., Songhori, E., Wang, S., Lee, Y.J., Johnson, E., Pathak, O., Bae, S., et al.: Chip placement with deep reinforcement learning. arXiv preprint arXiv:2004.10746 (2020)

73. Mitchell, S., Potash, E., Barocas, S., D'Amour, A., Lum, K.: Algorithmic fairness: Choices, assumptions, and definitions. Annual Review of Statistics and Its Application **8**, 141–163 (2021)

74. Moret-Bonillo, V.: Can artificial intelligence benefit from quantum computing? Progress in Artificial Intelligence **3**(2), 89–105 (2015)

75. Mueller-Gastell, J., Sena, M., Tan, C.Z.: A multi-digit OCR system for historical records (computer vision)

76. Nagaraj, A., Kejriwal, M.: Robust quantification of gender disparity in pre-modern English literature using natural language processing. CoRR **abs/2204.05872** (2022). DOI URL https://doi.org/10.48550/arXiv.2204.05872

77. Nam, D., Kejriwal, M.: How do organizations publish semantic markup? three case studies using public schema.org crawls. Computer **51**(6), 42–51 (2018). DOI URL https://doi.org/10.1109/MC.2018.2701635

78. Pearl, J.: Causal inference. NIPS Causality: Objectives and Assessment pp. 39–58 (2010)

79. Pompili, M., Hermans, S.L., Baier, S., Beukers, H.K., Humphreys, P.C., Schouten, R.N., Vermeulen, R.F., Tiggelman, M.J., dos Santos Martins, L., Dirkse, B., et al.: Realization of a multinode quantum network of remote solid-state qubits. Science **372**(6539), 259–264 (2021)

80. Rakocevic, G., Djukic, T., Filipovic, N., Milutinović, V.: Computational medicine in data mining and modeling. Springer (2013)

81. Robertson, A.: The US copyright office says an AI can't copyright its art (2022). URL https://www.theverge.com/2022/2/21/22944335/us-copyright-office-reject-ai-generated-art-recent-entrance-to-paradise

82. Schmidt, J., Johnson, C., Eason, J., MacLeod, R.: Applications of automatic mesh generation and adaptive methods in computational medicine. In: Modeling, Mesh Generation, and Adaptive Numerical Methods for Partial Differential Equations, pp. 367–393. Springer (1995)

83. Selvam, R.K., Kejriwal, M.: On using product-specific schema.org from web data commons: An empirical set of best practices. CoRR **abs/2007.13829** (2020). URL https://arxiv.org/abs/2007.13829

84. Selvaraju, R.R., Cogswell, M., Das, A., Vedantam, R., Parikh, D., Batra, D.: Grad-cam: Visual explanations from deep networks via gradient-based localization. In: Proceedings of the IEEE international conference on computer vision, pp. 618–626 (2017)

85. Sidhu, S.: Unicef calls for averting a lost generation as covid-19 threatens to cause irreversible harm to children's education, nutrition and well-being. (2020). URL https://www.unicef.org/press-releases/unicef-calls-averting-lost-generation-covid-19-threatens-cause-irreversible-harm

86. Sironi, C.F.: Monte-Carlo tree search for artificial general intelligence in games (2019)

87. St-Hilaire, F., Vu, D.D., Frau, A., Burns, N., Faraji, F., Potochny, J., Robert, S., Roussel, A., Zheng, S., Glazier, T., et al.: A new era: Intelligent tutoring systems will transform online learning for millions. arXiv preprint arXiv:2203.03724 (2022)

88. Sundararajan, M., Najmi, A.: The many Shapley values for model explanation. In: International conference on machine learning, pp. 9269–9278. PMLR (2020)

89. Swire, B., Berinsky, A.J., Lewandowsky, S., Ecker, U.K.: Processing political misinformation: Comprehending the trump phenomenon. Royal Society open science **4**(3), 160802 (2017)

90. Tarrant, A., Cowen, T.: Big Tech Lobbying in the EU. The Political Quarterly, **93**: 218–226 (2022). https://doi.org/10.1111/1467-923X.13127

91. Taskin, L., Al Amoudi, I.: Humanizing management: Foundation, precautions and prospects. In: 36th European Group for Organization Studies (EGOS) annual meeting (2020)

92. Taylor, C.A., Draney, M.T., Ku, J.P., Parker, D., Steele, B.N., Wang, K., Zarins, C.K.: Predictive medicine: computational techniques in therapeutic decision-making. Computer Aided Surgery: Official Journal of the International Society for Computer Aided Surgery (ISCAS) **4**(5), 231–247 (1999)

93. Temir, E.: Deepfake: new era in the age of disinformation & end of reliable journalism. Selçuk İletişim **13**(2), 1009–1024 (2020)

94. Tomašev, N., Cornebise, J., Hutter, F., Mohamed, S., Picciariello, A., Connelly, B., Belgrave, D., Ezer, D., Haert, F.C.v.d., Mugisha, F., et al.: AI for social good: unlocking the opportunity for positive impact. Nature Communications **11**(1), 1–6 (2020)

95. Vallor, S.: Artificial intelligence and public trust (2017)

96. Wang, J., Hu, X.: Gated recurrent convolution neural network for OCR. Advances in Neural Information Processing Systems **30** (2017)

97. Wang, K., Gou, C., Duan, Y., Lin, Y., Zheng, X., Wang, F.Y.: Generative adversarial networks: introduction and outlook. IEEE/CAA Journal of Automatica Sinica **4**(4), 588–598 (2017)

98. Wei, T.C., Sheikh, U., Ab Rahman, A.A.H.: Improved optical character recognition with deep neural network. In: 2018 IEEE 14th International Colloquium on Signal Processing & Its Applications (CSPA), pp. 245–249. IEEE (2018)

99. Westerlund, M.: The emergence of deepfake technology: A review. Technology Innovation Management Review **9**(11) (2019)

100. Wills, K.: Ai around the world: Intellectual property law considerations and beyond. J. Pat. & Trademark Off. Soc'y **102**, 186 (2021)

101. Winslow, R.L., Trayanova, N., Geman, D., Miller, M.I.: Computational medicine: translating models to clinical care. Science translational medicine **4**(158), 158rv11– (2012)

102. Yeung, K., Lodge, M.: Algorithmic regulation. Oxford University Press (2019)

103. Ying, M.: Quantum computation, quantum theory and ai. Artificial Intelligence **174**(2), 162–176 (2010)

104. Zhang, S., Kejriwal, M.: Concept drift in bias and sensationalism detection: an experimental study. In: F. Spezzano, W. Chen, X. Xiao (eds.) ASONAM '19: International Conference on Advances in Social Networks Analysis and Mining, Vancouver, British Columbia, Canada, 27–30 August, 2019, pp. 601–604. ACM (2019). DOI URL https://doi.org/10.1145/3341161.3343690

105. Zhang, Y., Song, K., Sun, Y., Tan, S., Udell, M.: "Why should you trust my explanation?" understanding uncertainty in lime explanations. arXiv preprint arXiv:1904.12991 (2019)

106. Zhou, Z., Kejriwal, M., Miikkulainen, R.: Extended scaled neural predictor for improved branch prediction. In: The 2013 International Joint Conference on Neural Networks, IJCNN 2013, Dallas, TX, USA, August 4-9, 2013, pp. 1–7. IEEE (2013). DOI URL https://doi.org/10.1109/IJCNN.2013.6707059

Glossary

Artificial Intelligence (AI)   While the term "AI" dates back to the 1950s, there is still a lack of a single agreed-upon definition, although the prevailing definitions are qualitatively similar. One such definition, from a pragmatic source geared toward industry [1], that we choose to adopt here is that it is the "ability of computers to perform cognitive functions associated with human minds, such as perceiving, reasoning, learning, and problem solving." While not explicit in the definition, the ability is assumed to derive from a model that is encoded in software (even if it eventually used to control hardware, such as actuators in a robot). For example, a computer vision model that can automatically recognize (or "perceive") faces in images is a program that has usually been trained on a large number of labeled faces, and that can be used on a device (even a camera), once trained, to recognize faces in images that it has not seen during training.

Supervised Machine Learning   This branch of machine learning involves giving machines the ability to learn a function from labeled training data, with the aim of predicting the label when given test data. For example, given tweets that are labeled as either positive or negative sentiment, the machine would learn a *classifier* that would be able to classify tweets it had not seen before as having positive or negative sentiment. In this example, the label is categorical (hence, the function is called a classifier), but it could also be real (in which case, the function would be a regressor, the simplest example of which is ordinary least-squares linear regression). Over the decades, many different learning algorithms and architectures have been proposed for supervised machine learning, including logistic regression, decision trees, random forests, and artificial neural networks. With recent advances in deep learning, the last of these has emerged as a clear winner, since it is able to learn highly non-linear and complex surfaces given adequate amounts of labeled data.

Unsupervised Machine Learning   This branch of machine learning is similar to supervised machine learning in that a sample of datapoints must be given to the machine, but the key difference is that the data is not labeled. Instead, the machine's task is to discover structure over the data, using techniques such as clustering. A practical use case of unsupervised machine learning is visualization. Representation learning, which is subsequently described, also has

© The Author(s), under exclusive license to Springer Nature Switzerland AG 2023

M. Kejriwal, *Artificial Intelligence for Industries of the Future*, Future of Business and Finance, https://doi.org/10.1007/978-3-031-19039-1

interesting connections to unsupervised machine learning. It should be noted that hybrid versions of unsupervised machine learning, containing some elements of supervised machine learning, have also been proposed in some application contexts; hence, the line between the two is not always as crisp as one might imagine. Machine learning paradigms such as self-supervised learning, transfer learning and active learning do not neatly fit into one or the other, since they do involve some supervision, but it is very minimal compared to traditional supervised machine learning. One can therefore imagine "pure" supervised and unsupervised machine learning as lying at the extremes of a spectrum, with some of these other paradigms occupying points along the spectrum.

Reinforcement Learning   This branch of machine learning is inspired by behaviorism in psychology, and aims to learn a policy by exploring the environment and action-space, while also seeking to "exploit" actions that have previously yielded a "reward" from the environment. It is especially appropriate for gameplaying and other such environments where there is no natural concept of training data (unlike supervised machine learning) or structure discovery (unlike unsupervised machine learning) but there is an objective notion of how successful the policy is. For example, policies that lead to an agent winning the game more often than not would be considered to be better than those that led to an agent losing the game more often. Rewards in reinforcement learning can sometimes be sparse, with a single win/loss signal being provided at the end of the game. In recent years, *deep* reinforcement learning (which uses deep neural networks to learn a reinforcement learning policy), developed and refined by companies like DeepMind and evaluated against human champions on games like Chess and Go, have met with enormous success.

Deep Learning   Although artificial neural networks (ANNs), inspired loosely by the biological organization of the human brain, were proposed many decades ago and the science on them has considerably evolved, their true potential was (arguably) unveiled because of the empirical success achieved by deep learning since the early 2010s. Deep learning is usually synonymous with the training and optimization of so-called *deep neural networks*, which are ANNs with certain structural properties, such as the presence of many layers of neurons. As the field has advanced, multiple types of deep neural networks have been proposed and studied, including convolutional neural networks (CNNs), graph neural networks (GNNs) and transformers. Each of these seems to have specific strengths, although there is still much left to be understood about the individual merits of these networks. In practice, CNNs have proven to be enormously successful for computer vision tasks, while transformers are now dominant in sequence-oriented tasks, which are common in natural language processing. In general, however, deep learning systems have emerged as state-of-the-art in numerous subfields of AI, owing to their mathematical ability to automatically learn highly complex representations (see below) of large, messy real-world datasets, and to learn non-linear surfaces that allow for more accurate and powerful supervised machine learning.

Representation Learning   Representation learning is an important sub-task in modern machine learning research where neural networks, or other similar technologies, are used to automatically learn "representations" of raw data, such as images and text. Typically, these representations are real-valued vectors that have convenient properties, e.g., they may be low-dimensional, and often take context into account. Improvements in representation learning, compared to the previous generation of manual feature engineering and selection methods, are often cited as the underlying reason why neural networks are able to work so effectively with large quantities of noisy, high-dimensional data.

Augmented Artificial Intelligence   In contrast with "fully automated" AI, where the goal is to automate away the capabilities of a human being (whether in the context of a narrow task, a job, or a suite of functions), augmented AI attempts to augment the human's capability by automating aspects of the task that are prone to continuous repetition and human error. In an ideal world, this would lead to a person becoming more productive (and happier) by focusing on the aspects of the job where the marginal utility of their unique skills and creative talents can be optimally deployed.

Natural Language Processing   Also known as natural language *understanding*, natural language processing (NLP) is one of the oldest and most important subfields of AI that is concerned with giving a machine the ability to understand human language as it is spoken and written. Human languages in this context are associated with the umbrella term "natural language." The practical focus tends to be on written, rather than spoken, natural language "text," with a different line of research (called speech recognition) generally applicable when audio is the direct input. NLP tasks include information extraction, question answering, and text summarization, to only name a few. In recent years, the advent of transformer-based models such as BERT [2] and GPT-3 [3] have led to superlatively high performance on some of these tasks.

Knowledge Graphs   A knowledge graph (KG) is a structured representation of domain (or more broadly, encyclopedic) knowledge where entities and attributes are vertices in a graph, and relationships connecting pairs of entities, or attributes to entities, are directed edges in the graph. Hence, both vertices and edges have labels. The semantics and constraints of the knowledge in the KG is typically defined by an ontology. Standard ontologies already exist in many domains, e.g., the Gene Ontology in the biological domain.

Language Representation Models   Technically, a *language model* is defined as a probability distribution over sequences of words. Language representation models (LRMs), generally synonymous with *neural* language models, are based on learning continuous real-valued vector representations of language units, such as words and sentences, using deep learning techniques (transformers being the state-of-the-art in the NLP community). In its modern form, an LRM is first *pre-trained* on large quantities of text, such as Wikipedia, the Google News corpus and other such natural language texts. Because this is an expensive process, especially for the recent batch of transformers, pre-training is often conducted in industry or by organizations such as OpenAI. Some models, such as BERT, are publicly available, while others (such as GPT-3) are available using a paid API. To be applicable to specific problems, such as information extraction, an LRM has to be *fine-tuned*. This process requires specialized task-specific data, but the data is much smaller in size than pre-trained data, and the process takes a lot less time. Researchers and other stakeholders can therefore download the pre-trained models and fine-tune them on their own datasets to achieve a customized high-performance model. LRMs have been getting bigger over time, with trillion-parameter models expected to become common (or at least in-use) within the current decade. Some have criticized these models on account of the cost of training them, and their impact on the climate due to the energy-intensive compute required. However, the dramatic improvements in task-performance often observed with increase in size are pragmatically difficult to argue against, especially given the pace of competition in the tech industry.

Dataset Bias   This is a type of bias that is related to the problem, of selection bias in the social sciences. The bias arises when datasets are not necessarily representative of the overall population (e.g., an image dataset may be under-represented in terms of the numbers of people of color, or non-male genders, present in it). AI, and particularly, deep learning, systems trained on such datasets will exhibit problems when deployed in the real world. For example, a facial recognition system may make more errors when shown pictures of people of color. Dataset bias can also lead to other statistical issues, including lack of generalization, as some researchers have attempted to show in the computer vision community (in particular) [4]. The extent of the bias in other AI domains is not completely evident, even today.

Global Data Protection Regulation (GDPR)   GDPR is a regulation in the European Union (EU) with personal data at its core, and a cornerstone example of a set of seven guiding principles spanning multiple nations, and with some moderate success already in regulating Big Tech. Importantly, businesses based outside the EU, but doing business in the EU or collecting data on EU citizens, are also subject to the GDPR. GDPR is important to AI regulation as well, given that modern AI (such as deep neural networks) is famously data hungry; additionally, the same companies that are developing and deploying many of these AI models also tend to rely most on advertising revenue and personalized consumer targeting, which makes these companies natural targets of GDPR regulation.

Deepfakes   Deepfakes are a novel application of AI wherein neural networks are used to generate "fake," but highly realistic, images and videos of real individuals. Deepfakes are now regulated in many parts of the world, and there is significant evidence that they are being used for malicious purposes such as spreading misinformation, conspiracy theories, and extremist propaganda. At the same time, beneficial and creative uses of the technology have also been proposed, providing a case study of the old adage that technology itself is not inherently good or bad.

Artificial General Intelligence   An original goal of early Artificial Intelligence research, Artificial General Intelligence is the ability of a single machine or architecture to perform at a human level on several important tasks, rather than just one, and to be able to make contextual decisions. It is not completely settled what the specific range of tasks is that needs to be solved, and at what precise performance, for this to be achieved. However, some cases are clearer than others. For example, a robot that would be capable of doing several household chores and having a natural conversation with the (human) resident is further along the path than a customer service chatbot that is only capable of responding to a narrowly tailored range of requests.

References

1. McKinsey: An executive's guide to AI (2020). URL https://www.mckinsey.com/business-functions/quantumblack/our-insights/an-executives-guide-to-ai
2. Devlin, J., Chang, M.W., Lee, K., Toutanova, K.: Bert: Pre-training of deep bidirectional transformers for language understanding. arXiv preprint arXiv:1810.04805 (2018)
3. Brown, T.B., Mann, B., Ryder, N., Subbiah, M., Kaplan, J., Dhariwal, P., Neelakantan, A., Shyam, P., Sastry, G., Askell, A., Agarwal, S., Herbert-Voss, A., Krueger, G., Henighan, T., Child, R., Ramesh, A., Ziegler, D.M., Wu, J., Winter, C., Hesse, C., Chen, M., Sigler, E., Litwin, M., Gray, S., Chess, B., Clark, J., Berner, C., McCandlish, S., Radford, A., Sutskever, I., Amodei, D.: Language models are few-shot learners. CoRR **abs/2005.14165** (2020). URL https://arxiv.org/abs/2005.14165
4. Torralba, A., Efros, A.A.: Unbiased look at dataset bias. In: CVPR 2011, pp. 1521–1528. IEEE (2011)

Index

© The Author(s), under exclusive license to Springer Nature Switzerland AG 2023
M. Kejriwal, *Artificial Intelligence for Industries of the Future*, Future of Business
and Finance, https://doi.org/10.1007/978-3-031-19039-1

The manufacturer's authorised representative in the EU is Springer
Nature Customer Service Centre GmbH, Europaplatz 3, 69115 Heidelberg,
Germany. If you have any concerns regarding our products, please
contact ProductSafety@springernature.com

Printed and bound by CPI Group (UK) Ltd, Croydon, CR0 4YY

29/04/2026

02099527-0005